Introduction to
Imaging from Scattered Fields

Introduction to
Imaging from Scattered Fields

Michael A. Fiddy
University of North Carolina at Charlotte

R. Shane Ritter
Olivet Nazarene University
Bourbonnais, Illinois

CRC Press
Taylor & Francis Group
Boca Raton London New York

CRC Press is an imprint of the
Taylor & Francis Group, an **informa** business

MATLAB® is a trademark of The MathWorks, Inc. and is used with permission. The MathWorks does not warrant the accuracy of the text or exercises in this book. This book's use or discussion of MATLAB® software or related products does not constitute endorsement or sponsorship by The MathWorks of a particular pedagogical approach or particular use of the MATLAB® software.

CRC Press
Taylor & Francis Group
6000 Broken Sound Parkway NW, Suite 300
Boca Raton, FL 33487-2742

First issued in paperback 2019

© 2015 by Taylor & Francis Group, LLC
CRC Press is an imprint of Taylor & Francis Group, an Informa business

No claim to original U.S. Government works

ISBN-13: 978-1-4665-6958-4 (hbk)
ISBN-13: 978-0-367-86775-1 (pbk)

**Visit the Taylor & Francis Web site at
http://www.taylorandfrancis.com**

**and the CRC Press Web site at
http://www.crcpress.com**

This book is dedicated to my wife, Carolyn, and our family,
Howard, Aidan, Sam, Colleen, Grant, and Rosa.
Michael A. Fiddy

This book is dedicated to my wife, Julie, and
our family, Kayla, David, Matthew, Ruth, Sarah,
Susanna, Daniel, Aaron, Abbie, and Micaiah.
R. Shane Ritter

Contents

SECTION IV   Appendices

Preface

The objective of this book is to present an overview of the challenging problem of determining information about an object from measurements of the field scattered from that object. This problem is a very old one, since, in a fundamental sense, most of what we perceive and learn about objects around us is a result of electromagnetic or acoustic waves impinging on, interacting with and scattering from those objects. The theoretical formalism of a scattering problem is increasingly complex, as the extent of the interactions increase between the fields with the object. The forward or direct problem generally demands a good model for the anticipated response of the object. Deducing information about the object generally demands knowledge of that model or that (acceptable) approximations can be made to simplify matters. Theoretical approaches to solving inverse problems have been widely studied and as a broad class of problems are known to suffer from concerns over lack of uniqueness and solution stability (ill-conditioning) but, despite modeling a physically well-defined problem, could also be formulated in a way that the very existence of a solution is questionable. In the specific context of inverse scattering theories and algorithms, we present in this text an overview of some of the more widely used approaches to recover information about objects. We consider both the assumptions made *a priori* about the object as well as the consequences of having to recover object information from limited numbers of noisy measurements of the scattered fields.

There is a wealth of literature dealing with scattering and inverse scattering methods for relatively simple structures embedded in a homogeneous background. We introduce the terminology and concepts early in the text and review some important inverse methods. When the scattering is assumed to be "weak," which we define in the text, inversion methods allow more straightforward inverse algorithms to be exploited. We highlight the consequences of the widespread practice of adopting such methods when they are not justified while recognizing their attractiveness from a practical implementation point of view. Assuming weak scattering allows many well-established techniques developed in Fourier-based signal and image processing to be incorporated. The weak scattering models facilitate a simple mapping of scattered field data onto a locus of points in the Fourier domain of the object of interest. More rigorous scattering methods that rely on iterative techniques or strong prior knowledge of the forward scattering model are often slow to implement and may not yield reliable information.

Over the last several years, we have been developing and improving an approach which, while governed by the usual limitations associated with inverse problems, retains many advantages in terms of implementing the weak scattering methods while addressing directly the multiple and strongly scattering phenomena that occur with most objects of interest. The approach is based on a nonlinear filtering step in the inverse algorithm, which requires some preprocessing of the measured data. We illustrate how one can use this algorithm in a very practical way, providing MATLAB® code to help quickly begin applying the approach to a wide variety of inverse scattering problems. We illustrate it using a number of two-dimensional electromagnetic scattering examples.

In later chapters of the book, we draw attention to some very important and often forgotten overarching constraints associated with exploiting inverse scattering algorithms. The inherent lack of uniqueness of a solution to an inverse problem when using finite data requires either implicitly or explicitly that a single solution be selected somehow. A figure of merit or cost function is needed to restore some confidence to the interpretation of the calculated image of the scattering object. The number of measurements made has an obvious and very significant effect on the quality and reliability of an object reconstruction. We explain how considerations of the number of degrees of freedom associated with any given scattering experiment can be found and how this dictates a minimum number of data that should be measured. We argue that estimating the properties of an object from scattered field measurements necessarily requires some prior estimate of the volume from which scattered field data are collected. The use of prior knowledge about the object or properties of the illuminating fields can be used for this purpose to good effect. We describe in detail what we refer to as the prior discrete Fourier transform or "PDFT" algorithm, which accomplishes this. The PDFT restores stability and improves estimates of the object even with severely limited data, provided it is sufficient to meet a criterion based on the number of degrees of freedom.

We have organized this book with graduate students and those practicing imaging from scattered fields in mind. This includes, for example, those working with medical, geophysical, defense, and industrial inspection inverse problems. It will be helpful for readers to have an understanding of basic electromagnetic principles, some background in calculus and Fourier analysis, and preferably familiarity with MATLAB (and possibly COMSOL®) in order to take advantage of the source code provided. The text is self-contained and gives the required background theory to be able to design improved experiments and process measured data more effectively, to recover for a strongly scattering object an estimate that is not perfect, but probably the best that one can hope for from limited scattered field data.

The authors would like to acknowledge their productive collaborations over the years on imaging and inverse scattering with Umer Shahid, Charlie Byrne, Markus Testorf, Bob McGahan, and Freeman Lin.

MATLAB® is registered trademark of The MathWorks, Inc. For product information, please contact:

The MathWorks, Inc.
3 Apple Hill Drive
Natick, MA 01760-2098 USA
Tel: 508 647 7000
Fax: 508-647-7001
E-mail: info@mathworks.com
Web: www.mathworks.com

Supplementary materials including MATLAB code for exercises are available on the book's page at www.crcpress.com. Please visit the site, look up the book and click to the Downloads and Updates tab.

Authors

Michael A. Fiddy received the PhD degree from the University of London in 1977 and was a research fellow in the Department of Electronic and Electrical Engineering at the University College London before becoming a faculty member at the London University (King's College) in 1979. He moved to the University of Massachusetts Lowell in 1987 where he was Head, Department of Electrical and Computer Engineering from 1994 until 2001. In January 2002, he was appointed the founding director of the newly created Center for Optoelectronics and Optical Communications at The University of North Carolina at Charlotte. He has been a visiting professor at the Institute of Optics, Rochester, NY; Mathematics Department, Catholic University, Washington, DC; Nanophotonics Laboratory, Nanyang Technical University, Singapore; and Department of Electrical and Computer Engineering, University of Christchurch, Christchurch, New Zealand. He has also been the Editor-in-Chief of the journal *Waves in Random and Complex Media* since 1996 and holds editorial positions with several other academic journals. He was the topical editor for signal and image processing for the *Journal of the Optical Society of America* from 1994 until 2001. He has chaired 20 conferences in his field, and is a fellow of the OSA, IOP, and SPIE. His current research interests are inverse problems related to super-resolution and meta-material design.

R. Shane Ritter holds a BS and MS in electrical engineering from Mississippi State University and a PhD in electrical engineering from the University of North Carolina at Charlotte. He is currently the chair of the Engineering Department in the School of Professional Studies at Olivet Nazarene University, Bourbonnais, IL, where he also serves as a professor of electrical and computer engineering. He has also served as the director of electrical engineering for a number of engineering firms, as well as an independent consulting electrical engineer in many different aspects of electrical engineering. He is currently licensed as a professional engineer in over 35 states and is also a Registered Communication Distribution Designer (RCDD). Shane served as an adjunct faculty member in mathematics, statistics, and research at the University of Phoenix from 2001 until 2009. Shane also served as an adjunct faculty member in electrical and electronics engineering at the ITT Technical Institute in Charlotte, NC in 2010.

List of Symbols

CHAPTER 1

e	Euler's constant (2.71828182845905)
$f(x,y)$	Inverse Fourier transform
$F(x,y)$	Fourier transform
k	Wave number ($=2\pi/\lambda$)
k_x, k_y	Spatial frequency
$\mathbf{r}_{inc}$	Radius vector in direction of incident wave
$\mathbf{r}_{sct}$	Radius vector in direction of the scattered wave
$\mathbf{R}$	Radius vector
$u(t,s)$	System response
V_m	Max or mean of $V(\mathbf{r})$
$V(\mathbf{r})$	Target in terms of $\mathbf{r}$
δ	Delta function
∇	Gradient operator
Λ	Wavelength
π	pi (3.14159265358979)
ϕ	Phase angle
Ψ_s	Scattered field
Ψ_s^{BA}	Scattered field in the Born approximation

CHAPTER 2

B	Magnetic induction
c_0	Speed of light in a vacuum ($= (\varepsilon_0 \mu_0)^{1/2}$)
D	Electric displacement
E	Electric field
E_0	Electric field for incident wave
$E_0(\omega)$	Complex electric amplitude
G	Green's function
H	Magnetic field
$H_0(\omega)$	Complex magnetic amplitude
I	Unit tensor
J	Free current density
J_c	Conduction current density
J_s	Source current density
J_{su}	Source current density
M	Magnetization
n	Refractive index ($= (\varepsilon_r \mu_r)^{1/2}$)
$\hat{n}$	Unit vector normal to the interface pointing from the input medium into the second medium
P	Polarization
q	Dipole moment

V	Volume
ε_0	Electric permittivity
μ_0	Magnetic permeability
μ_r	Relative magnetic permeability $(1 + \chi_m)$
ρ	Free charge density
ρ_{su}	Surface charge density
χ_e	Electric susceptibility
χ_m	Magnetic susceptibility

CHAPTER 4

a	Measure of physical size of target
$f(k\hat{r}, k\hat{r}_{inc})$	Scattering amplitude
$f^{BA}(k\hat{r}, k\hat{r}_{inc})$	Scattering amplitude in the Born approximation
$G_0(r, r')$	Green's function in free space
$H_0^{(1)}$	Zero order Hankel function of the first kind
$\hat{r}_{inc}$	Unit vector that specifies the direction of the incident field
ε_0	Permittivity of free space
$\varepsilon(r)$	Target permittivity as a function of radius
ϕ_{inc}	Angle of incident wave with the x-axis
ϕ_s	Angle of scattered wave with x-axis
$\Phi(r)$	Complex phase function
$\Phi_s(r)$	Complex phase function of the scattered field
$\Psi(r)$	Total field in terms of r
Ψ_s	Scattered field
Ψ_{inc}	Solution for the incident wave
Ω	Solid angle over H^2

CHAPTER 6

A_V	Target area
B_V	Target volume
n_{max}	Maximum index of refraction
$N_{2\text{-}D}$	Minimum degrees of freedom required in 2-D
$N_{3\text{-}D}$	Minimum degrees of freedom required in 3-D
Q	Mie measure of scattering cross section $(B_V n_{ma}/\lambda^2)$

CHAPTER 7

BIM	Born iterative method
CGM	Conjugate gradient method
D	Difference in simulated and measured fields
DBIM	Distorted Born iterative method
f_{PDFT}	PDFT estimator function
$p(r)$	Non-negative prior weighting function
P	Fourier transform of $p(r)$
RRE	Relative residual error
$V_{BA}^1(r)$	First estimate or starting point of the BIM

$X[V_{\text{est}}(\boldsymbol{r})]$	Norm of discrepancy between simulated and measured scattered fields
ξ	PDFT weighted error
τ	Regularization constant
$\boldsymbol{\Psi}_s^{\text{sim}}$	Simulated field
$\boldsymbol{\Psi}_s^{\text{measured}}$	Measured field

I

FUNDAMENTALS

One

Introduction to Inverse Scattering

1.1 INTRODUCTION

Considerable knowledge of the world around us is based on receiving and interpreting electromagnetic and acoustic waves. We extend the bandwidths and sensitivities of our senses by using instruments and collecting radiation from sources and scatterers of radiation. Active illumination or insonification of objects to probe and image their structures is an important tool in advancing our knowledge. However, we need to have a good physical model that describes the possible interactions of those waves with scattering objects. Constitutive parameters (such as permittivity, permeability, refractive index, impedance, etc.) that have spatially and temporally varying properties describe the scattering objects. Wave propagation and scattering characteristics are governed by the fundamental relationships between these properties and their effects on the components of the field, as governed, for example, by Maxwell's equations. In either the electromagnetic case or the acoustical case, we need to derive a wave equation, both in differential or integral form, with appropriate boundary conditions or coefficients, and then analytically or numerically solve that equation to find the field outside the object. This so-called "direct" problem, which assumes that the object parameters are known and scattered fields are to be determined, is itself a nontrivial exercise but well defined.

The complementary or "inverse" problem is much more difficult and is the focus of this book. Making measurements of the scattered field at various locations near or far from the object takes time and effort. One has to specify the incident field properties such as wavelength, polarization, and direction, and then, relative to these, measure the scattered field properties. The question immediately arises as to how many measurements does one need in order to recover the information one wants about the object being probed. Inverting the governing wave equation is, from a purely mathematical perspective, the so-called ill-posed problem. Such problems require that one formally establish the following:

1. Whether there is a solution at all.
2. Whether the solution, should it exist, is unique.
3. Whether a calculated solution is or is not ill conditioned.

In most practical situations, one only measures a finite number of data on the scattered field, and uniqueness is impossible. One can fit an infinite number of functions (i.e., images) to a finite data set. This lack of uniqueness requires that we either explicitly or implicitly adopt a uniqueness criterion such as minimum energy, maximum entropy or some other such criterion using which one can define a unique solution and hope that it has a physical meaning.

In some imaging applications, one cannot measure the scattered field itself, for example at very high frequencies. Above ~1 THz we do not have detectors fast enough to measure the fluctuating field and all we acquire is a time averaged quantity. In electromagnetic problems we assume this is proportional to the magnitude squared of the (electric) field, $|E|^2$. As we shall see, the information required to solve the inverse problem and calculate an image of the object demands that we solve another problem, namely that of estimating from $|E|$, the function $E = |E| \exp(i\phi)$ or solve the so-called phase-retrieval problem (ϕ denotes phase). This is also nontrivial and, without knowledge of the phase, the information we can recover about the object is severely limited and at best statistical in nature.

Most problematic is the inevitable presence of noise in our measured data. Inverse procedures, as we shall see in the coming chapters, are always ill conditioned. This means that one can expect small changes in the data as a result of noise to lead to very large differences in our images. The instability of inverse methods can be understood mathematically and remedied using the so-called regularization techniques. The price to be paid to control ill conditioning is a degradation of the image, for example, a loss of resolution. However, since we cannot guarantee a unique solution in practice, we have to accept further compromises in order to obtain an image we can have some confidence in.

From a practical standpoint, we hope to collect the minimal amount of data to provide the image quality needed for the task at hand. Maps of spatially varying contrast might suffice while, for other purposes, for example, in medical imaging, a quantitatively accurate map of a constitutive parameter such as impedance might be essential. Overarching all of these issues is the more important problem of the governing equation to be inverted being inherently nonlinear in nature. The scattered field for all but the weakest scattering objects depends on the complexity of the scattering processes that occur within the object itself. For inverse problems, for by very definition we do not know the structure of the object, we cannot know *a priori* the extent of multiple scattering that occurs within it. We can define what we mean by "weakly" scattering, and that assumption, while rarely valid in practice, does lead to a more tractable inversion method but one that still suffers from the questions of uniqueness and ill conditioning mentioned above. For more interesting, but strongly scattering objects, we need to address the nonlinear aspect of the inverse problem. We will describe methods that do this but emphasize now that there are, at the time of this writing, still no fast and reliable methods one can take off the shelf and use. Indeed, despite many decades of effort, inverse scattering methods remain very challenging and an active field of research. Methods we describe here have a range of applicability that limits their use to situations for which some prior knowledge about the object is available. This is certainly possible in some applications such as imaging a limb or probing a suitcase, and prior knowledge can play an important role in addressing the uniqueness question, as we shall see.

1.2 INVERSE SCATTERING PROBLEM OVERVIEW

The wavelength of the radiation used with respect to the scale of the features one wishes to image provides a convenient way to segregate inverse scattering problems. In the limit of the wavelength becoming relatively small, geometrical optics or ray-based approximations become reasonable. In the very high frequency limit, for example, when using x-rays, one can assume that the radiation emerging from an object has not been refracted at all, and the measured data are interpreted as a shadow of the attenuation in the object. The mathematics describing this is well established, dating back to Radon (1986). Johann Radon's original paper was published in 1917 (Radon, 1917). The Fourier transform plays an important role here and throughout this book (see Appendix A). The technique of computed tomography, which incorporates a Radon transform (Wolf, 1969), uses projection data which measures the line integral of an object parameter, for example, of $f(x,y)$ in the equation shown below, along straight lines (y-axis in this example). This enables the Fourier Slice Theorem to be used to build up information about $F(k_x,k_y)$ by rotating the illumination direction.

$$F(k_x,0) = \int_{-\infty}^{+\infty}\int_{-\infty}^{+\infty} f(x,y)e^{i(k_x x)}\,dx\,dy = \int_{-\infty}^{+\infty}\left\{\int_{-\infty}^{+\infty} f(x,y)dy\right\}e^{i(k_x x)}\,dx \qquad (1.1)$$

where k_x and k_y are the spatial frequency variables that have units of reciprocal distance, that is, $k_x x$ is dimensionless. When object constitutive parameter fluctuations or inhomogeneities such as refractive index fluctuations in a semitransparent object are comparable in size to the interrogating wavelength, then scattering or diffraction effects become significant. As we shall see, Fourier data on the object are still obtainable in this situation provided the Born or Rytov approximations are valid. We will describe these approximations, which allow inversion algorithms to be formulated, and we will discuss in detail the criteria for their validity. When inverting Fourier data there is the question of how to make the best use of the limited set of noisy samples available. At optical frequencies, there is also an additional problem: that the phase of the scattered field may only be measured with difficulty. Some methods for phase retrieval are discussed in Appendix B.

Usually, approximations are employed to make the scattered fields (which can be expressed by Fredholm integral equations of the first kind) more tractable for numerical computation. The merits of the Born and the Rytov approximations, and more sophisticated techniques derived from them, have spawned a lot of controversy over the years. A principle cause for controversy is that these approximations are based on the interpretation given when strong inequalities are met, in order to simplify (or linearize) the governing equation. The physical interpretation of imposing these inequalities can be rather subjective. It is also problematic that sometimes these approximations appear to provide reasonably good images when one might not expect them to. This issue also illustrates one of the cautionary messages to be conveyed when working with inverse problems, which is that deliberate or inadvertent inverse crimes can be committed! These are crimes by which, because of the difficulty of acquiring real data from known objects with which to test an inversion method, the direct problem is solved to generate data. Occasionally approximations made in solving the direct problem are the very ones employed in the inverse method, thereby increasing the chances that the recovered image will look good. Consequently, we spend some time in this book describing the importance of understanding the nature of scattered field data used to validate imaging algorithms and suggest methods to generate such data. This of course is only necessary in the absence of real measured data from known targets, but despite the best efforts of many, real data sets are still few and far between. Data provided since the early 1990s by the US Air Force and the Institut Fresnel (Belkebir and Saillard, 2001, 2005) have done a tremendous service in providing high quality data from known objects, which provides a means for comparing different inverse scattering techniques and thereby improves them. For a scatterer of compact support (with d as the size of its largest dimension), the qualitative statement is usually made that the (first) Born approximation is valid only when the scatterer is "small" on the scale of the incident wavelength; this is discussed in more detail in Chapter 4.

The Born series solution to the integral equation of scattering is an infinite series which is traditionally defined as only valid when the criterion $kV_{m}d < 1$ is met. Here, k is the wavenumber $k = 2\pi/\lambda$ where λ is a measure of the wavelength *inside* the scattering object. This is obviously difficult to determine for an unknown object's constitutive parameter $V(r)$ that is being imaged. V_{m} is some measure of the maximum or mean value of $V(r)$ which is also unknown; consequently there is a temptation to apply the first Born approximation. This requires that $kV_{m}d \ll 1$ and, as we shall see, makes recovering an image computationally straightforward. Indeed it reduces the inverse scattering problem to one of a limited-data Fourier estimation problem. This is a problem on which there is much written, and it provides a comfort zone in which to work and process scattered field data, in the (vain) hope that images obtained when $kV_{m}d$ is not less than 1 still convey some meaningful information. The criterion for the validity of the Rytov approximation is equally vague, relying on the qualitative statement that spatial fluctuations in V be slow on the scale of the wavelength, but that the magnitude of the fluctuations of V need not necessarily be small or of low contrast. In other words, this physical interpretation of the validity of the Rytov approximation is based on the requirement that the absolute value of the rate of change of the complex phase of the scattered field within V be small compared with $k\nabla V$, where ∇ is the gradient operator. If this assumption is reasonable, one can formulate the inverse Rytov method as a limited-data Fourier estimation problem as well.

An interesting and important question to ask is what errors are introduced if one does adopt the Born or Rytov approximation. This is a very reasonable and insightful step to take and doing so has revealed classes of objects for which one can expect the approximations to do poorly or fail altogether. There is also much to be said for bringing to the inverse scattering problem a wealth of signal and image-processing knowledge that has been established over the years for dealing with limited data, especially limited Fourier data. By more carefully formulating the inverse problem in terms of these approximations and having a description for the errors and artifacts the "first Born approximate" image might possess, one can develop methods to postprocess those images to try to recover $V(r)$. This is the approach we have adopted and will describe in more detail in a later chapter.

These inverse scattering algorithms that have been developed over the years, often referred to as diffraction tomography algorithms, fall into two classes. Devaney (1983) and Pan and Kak (1983) have modified the filtered back-projection algorithm used in conventional tomography to give a filtered back-propagation technique, and interpolation procedures have been studied (Kaveh and Soumekh, 1985) and both approaches compared when they were first proposed (Pan and Kak, 1983). Either procedure discussed above can be adopted when using the first Born approximation or Rytov approximation.

The basic formulation of inverse scattering is due to Wolf (1969) who studied the determination of object structure within the first Born approximation. The earliest and probably the best-known application of an inverse scattering theory in the first Born approximation was in experiments to study the structure of crystals and molecules using x-rays. Two major difficulties arise in such structure determination problems. The first is that the intensity of the scattering pattern of the object (or more accurately for a periodic structure, diffraction) is measured rather than the complex scattered field, and the second is that these intensity data are sampled at a rate determined by the reciprocal of the unit cell dimensions of the crystal. If the complex scattered field data were available over the entire (far field) scattering domain, then this rate of sampling would be reasonable according to Shannon's Sampling Theorem to adequately represent the scattered field. The intensity data should Fourier transform to give the autocorrelation of the unit cell, but the measured data are undersampled for this purpose, being sampled at the rate determined by the reciprocal of the unit cell rather than its autocorrelation function. In addition, physical constraints and signal-to-noise ratio realities mean that only a limited area of data is measurable. The consequences of this are that the recovery of object information is clearly severely ill posed requiring phase retrieval and both interpolation and extrapolation of the scattered field. It is for a good reason that many Nobel prizes were awarded for recovering the image of specific objects such as DNA or hemoglobin.

In practice, improved estimates of the scattered field can be made from this incomplete data set provided additional *a priori* information about the symmetry or structure of the unit cell is available. More recently, in the optical domain, use has been made of (quasi)monochromatic laser sources and holographic techniques to measure the complex scattered field. As was obvious to Wolf (1969), holograms store information about the three-dimensional (3-D) structure of a scattering object, and it should be possible to compute this structure from measurements performed on the hologram. The calculation of the amplitude and phase of the scattered field from measurements of the amplitude variations of the hologram was first described by Wolf and is a much more practical solution today than it was over 40 years ago. This added layer of difficulty is not ignored here, but our focus is on the inverse scattering procedures themselves and so we restrict our further discussions to lower frequency radiation sources. Clearly, even with the ultrasound waves in the megahertz region or microwave region, however, the maintenance of a precise phase reference while varying the directions of illumination might be a problem if it is not measured carefully. It is also difficult to measure all the data required to determine the 3-D structure of the objects, and most of the examples we show will be 2-D structures, not for any fundamental reason but purely for convenience of explanation and illustration.

1.3 DIFFRACTION TOMOGRAPHY

We have emphasized how the inverse scattering problem is the determination of physical parameters and features of an unknown object from a limited set of measurements of scattered field data. Future sections will go into this in more detail. When the object in question is strongly scattering, this is known well to be a very difficult problem to solve. While advances have been made primarily using iterative techniques, long computational times are necessary, and convergence is not guaranteed, especially if only incomplete and noisy data are measured. Most computationally feasible algorithms assume a weakly scattering condition in order to linearize the inversion procedure and can be broadly classified as diffraction tomography techniques. The primary motive for writing this book is to introduce a method, which remains relatively fast and simple to implement and which builds on these known inverse scattering algorithms (despite their limitations) and more advanced signal processing concepts. We have developed a method, which is simple to implement, and which can provide a good estimate of the image of the scatterer. It is a diffraction tomography technique that has built in to it a nonlinear filtering step, which directly addresses the limitations of assuming $kV_\mathrm{m}d \ll 1$ when it is clearly not correct. The success of the improved method inevitably depends on the quality and extent of the data, and it is useful if one has data from several illuminating wavelengths assuming that the constituent parameters representing the object do not change significantly over the frequency range employed. As we shall see, the quality of the reconstructed image depends on a number of factors. The new method is computationally fast and at the very least is useful in providing a quick look at the object features and serves as an initial guess for a more rigorous inversion scheme. Such a method based on diffraction tomography can obviously be used for both imaging and structure synthesis applications. If one measures the scattered field and can compute an image of the scattering object, then one can also specify a scattered field and thereby compute and define a scattering object. If one can make this object such that it will behave as expected, then this rich field of object synthesis or target camouflaging goes hand in hand with the imaging emphasis of this book. In the case of imaging from scattered field data, it is important that the content of the reconstructed image can be relied upon in order to discern features that might have some practical significance. The structure synthesis problem is not so constrained. In principle, any structure that provides the desired scattering characteristics is a solution to the synthesis problem. The only concern might be the ease with which the predicted scattering structure can be fabricated, that is, in terms of its material properties and structural features.

What is diffraction tomography? This family of inverse scattering techniques is formulated as a Fourier inversion procedure. Complex scattered field data are collected at a number of scattering angles in the (near or) far field for each angle of illumination as shown in Figure 1.1. As will be detailed in Chapter 4, the scattered far field data under these conditions, based on the first-order Born approximation, are mapped onto Ewald spheres in a space we will refer to as k-space which is the Fourier space associated with the real space in which the scattering object exists. With sufficient k-space coverage from measurements of scattered fields at different angles of incidence and different scattering angles, the inverse Fourier transform provides information about the scattering object (Figure 1.1). Traditionally, should $kV_\mathrm{m}d \ll 1$ then this information will be directly proportional to the object's permittivity or refractive index profile, relative to that of the background in which it resides.

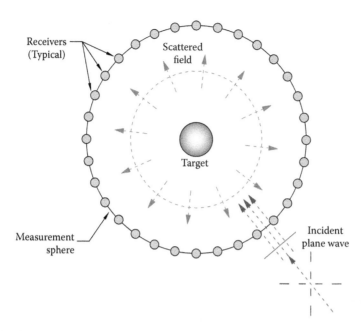

Figure 1.1 Typical experimental setup for diffraction tomography. The target or object is illuminated with a monochromatic plane wave and the scattered waves, after the interaction of the incident plane wave with the scattering object, are measured by receivers placed either in the near or far field around the object.

Of course the quality of that image will only be as good as the sampling of its Fourier transform in k-space. It is the limited sampling that inevitably results from experiments that leads to the need for considerable postprocessing and the application of estimation methods to get improved images. If the inequality is not appropriate, a distorted image results which may convey nothing useful whatsoever about the actual permittivity profile.

Ideally, the diffraction tomography method (Ritter, 2012), when restricted to weakly scattering targets, also assumes that the target is illuminated by a known quasi-monochromatic incident wave or waves, and the scattered field(s) is measured ALL around the target by a number of receivers. An illustration of a typical general experimental setup for this method is shown in Figure 1.2. The corresponding equation, which will be derived later in Chapter 4, is given by

$$\Psi_s^{BA}(\boldsymbol{r},\hat{\boldsymbol{r}}_{\text{inc}}) = \frac{1}{\sqrt{8\pi kr}}\, e^{i(kr+\pi/4)}k^2\int_D V(\boldsymbol{r}')e^{-ik(\hat{r}-\hat{r}_{\text{inc}})\mathbf{g}\boldsymbol{r}'}\,\mathrm{d}\boldsymbol{r} \qquad (1.2)$$

where Ψ_s is the scattered field and Ψ_s^{BA} is the scattered field in the Born approximation. It should be noted that $V(\boldsymbol{r})$ is assumed to be zero outside of the volume D. The relationship between the scattered field measurements and the scattering object is given by Equation 1.2. It can be seen that the Fourier variable in k-space is actually a locus of points mapping out a circle whose radius depends on the k-value of the incident wave and whose center depends on the direction of the incident plane wave. Deviating from incident plane waves and incident quasi-monochromatic waves complicates the interpretation of the scattered field using this Fourier picture quite considerably!

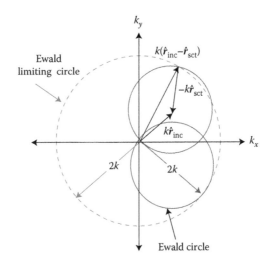

Figure 1.2 Fourier space (*k*-space) of the object as a result of interaction of different incident plane waves with scattering object. The direction of the incident field $\hat{r}_{inc}$ and the direction $\hat{r}_{sct}$ of a particular plane wave component of the scattered field define a point at the Ewald circle. Changing the incident field directions $\hat{r}_{sct}$ fills the interior of Ewald limiting circle.

From Equation 1.2, it can be seen that as k tends to infinity, corresponding to an almost zero wavelength, the radius of these Ewald circles becomes infinite, and for each incident wave direction, the circles become lines tangent to the k-space origin. This limiting case, when the Born approximation is valid, reduces to the Fourier Central Slice Theorem and the interpretation of projection data, for example, as used in x-ray CAT scanning that was mentioned earlier. This is a satisfying result but reinforces the fact that if $kV_m d \ll 1$ is not valid; because of multiple scattering, we should not expect to compute a useful image without taking further steps.

1.4 THEORETICAL ISSUES AND CONCERNS

In one dimension, the imaging from inverse scattering problem is well understood as discussed in detail in a paper by Zoughi (2000) and Devaney (1978). In general, in these 1-D problems, the scattered data to be inverted are treated as reflection and transmission coefficients. The inverse algorithm in these special cases is simply applied to this data to retrieve a reliable and unique 1-D profile of the original target.

The much more complicated 2-D version of this problem has been under examination for well over 100 years now. For the most part, there has been some limited success in implementing useful algorithms to perform this inversion for the very special case of weakly scattering targets. As already mentioned, the majority of these algorithms are based on using a technique to linearize this type of problem using either the Born approximation or the Rytov approximation, both of which will be discussed in more detail later. Historically, the success of these approaches or approximations is dependent on the target being a weak scatterer and that they fail to perform well when this is not the case. This historical viewpoint is due for a closer examination, as we shall see later.

A fundamental theoretical concern was mentioned earlier that we do not ignore (but also do not mention again): imaging from inverse scattered data is in general an ill-posed problem. By definition (Hadamard, 1923) a problem is considered ill posed if one (or more) of the following is true:

1. A solution does not exist.
2. A solution is nonunique.
3. The solution is unstable.

The first condition mentioned that a solution does not exist is definitely a possibility for these types of problems. For this discussion, it is assumed that a solution does exist, at least for the class of targets under consideration, because if the converse were true, further work would be pointless. There could be classes of targets where condition (i) above would be terminal. Even if a solution does exist, it is highly possible, albeit probable, that the solution is not unique. This means that two different sampled targets could produce the same scattered field patterns for a finite number of receivers. This being the case, the solution must be chosen from a solution space of possible solutions utilizing some global minimum, which could be problematic. This uniqueness problem could be amplified if the number of receivers is significantly low, which in a sense would lower the available degrees of freedom. The issue of degrees of freedom, its definition for this type of problem, and its effects will be examined more in Chapter 6.

The final "ill-posed" condition is that of stability (or a lack thereof). The stability of any inverse problem is a direct function of the system response as defined in Figure 1.3, where $y = hx$. In general, the inverse problem is stable if and only if h^{-1} exists and is stable. It is known that if h is continuous and if h^{-1} exists, then h^{-1} is also continuous. This in itself is not necessarily enough to ensure stability in general since the noise collected in performing measurements may also lead to instabilities and discontinuities being inserted in the valid data. It has been shown that, in mathematical theory and treatment for these types of problems, h can be represented by an integral equation which leads to $y = hx$ being a Fredholm integral equation of the first kind having a square integral (Hilbert–Schmidt) (Boas, 2011) kernel which in its most general form is written as

$$y(t) = hx(s) \quad y(t) = \int_a^b u(t,s)x(s)\,\mathrm{d}s \tag{1.3}$$

where $y(t)$ is the system output, $u(t, s)$ is the system response, and $x(s)$ is the object function or scatterer. This means that a minute error introduced in the measured data could quite possibly introduce a rather large error in the reconstructed results. For instance, if a solution x is perturbed by a delta function of the form

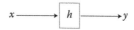

Figure 1.3   Definition of the system variables.

$$\delta x(s) = \varepsilon \sin(2\pi cs) \qquad (1.4)$$

where $c = 1, 2, 3, \ldots$ and ε is a constant. Then, the resulting perturbation in the output $y(t)$ would be given as

$$\delta y(t) = \varepsilon \int_a^b u(t,s)\sin(2\pi cs)\mathrm{d}s \qquad (1.5)$$

where $c = 1, 2, 3, \ldots$. Now if the Riemann–Lebesgue lemma (Renardy and Rogers, 2004) is applied, it follows that

$$\delta y \to 0 \quad as \quad c \to \infty$$

Now, if the integer "c" becomes large enough, the term $\|\delta x\|/\|\delta y\|$ can similarly become large as well, which results in Equation 1.3 becoming discontinuous over the domain of interest bringing us back to condition (iii) of an ill-posed problem. This situation can of course be controlled somewhat by taking precautions to reduce noise in the measurements and by strategically selecting the sample points. This is also complicated by the fact that there is a nonlinear relationship between the scatterer and the scattered field which can also make it very challenging to find a closed form solution for the 2-D inverse scattering problem. It is true that there are various approximations that have been developed to linearize these types of problems, but these techniques are by definition only valid for a limited class of targets.

One additional approach to this type of problem which will be examined in detail later is the use of nonlinear filtering techniques based upon homomorphic filtering to address the strong(er) scattering cases in general. This method is used in conjunction with diffraction tomography and the Born approximation to recover meaningful images of strongly scattering targets. As will be shown later, in applying this approach, the data are preprocessed to ensure that they are causal and minimum phase.

REFERENCES

Belkebir, K. and Saillard, M. 2001. Special section on testing inversion algorithms against experimental data, *Inverse Problems*, 17, 1565–1571.

Belkebir, K. and Saillard, M. 2005. Special section on testing inversion algorithms against experimental data: Inhomogeneous targets, *Inverse Problems*, 21, S1–S3.

Boas, R. P. 2011. *Entire Functions* (2nd edition). New York: Academic Press.

Devaney, A. J. 1978. Nonuniqueness in the inverse scattering problem. *Journal of Mathematical Physics*, *19*, 1526–1535.

Devaney, A. J. 1983. A computer simulation of diffraction tomography. *IEEE Transactions in Biomedical Engineering*, *30*, 377–386.

Hadamard, J. 1923. *Lectures on Cauchy's Problem in Linear Partial Differential Equations* (Dover Phoenix edition). New York: Dover Publications.

Kaveh, M. and Soumekh, M. 1985. Algorithms and error analysis for diffraction tomography using the Born and Rytov approximations. In Boerner W.-M. et al.

(editors), *Inverse Methods in Electromagnetic Imaging*, Part 2. Amsterdam: Springer, pp. 1137–1146.

Pan, S. X. and Kak, A. C. 1983. A computational study of reconstruction algorithms for diffraction tomography: Interpolation vs. filtered-backpropoagation. *IEEE Transactions on Acoustical Speech Signal Processing*, 31, 1262–1275.

Radon, J. 1986. On the determination of functions from their integral values along certain manifolds. *IEEE Transactions on Medical Imaging*, 5(4), 170–176.

Radon, J. 1917. Über die Bestimmung von FunktionendurchihreIntegralwerteläng sgewisserMannigfaltigkeiten, *Berichteüber die Verhandlungen der Königlich-SächsischenAkademie der Wissenschaftenzu Leipzig, Mathematisch-PhysischeKlasse* [Reports on the proceedings of the Royal Saxonian Academy of Sciences at Leipzig, mathematical and physical section], (69), 262–277, Leipzig: Teubner.

Renardy, M. and Rogers, R. C. 2004. *An Introduction to Partial Differential Equations, Texts in Applied Mathematics* (2nd edition, Vol. 13). New York: Springer-Verlag.

Ritter, R. S. 2012. *Signal Processing Based Method for Modeling and Solving Inverse Scattering Problems*. University of North Carolina at Charlotte, Charlotte: UMI/ProQuest LLC.

Wolf, E. 1969. Three-dimensional structure determination of semi-transparent objects from holographica data. *Optics Communication*, 1, 153–156.

Zoughi, R. 2000. *Microwave Nondestructive Testing and Evaluation*. Amsterdam: Kluwer Academic Publishers.

Two

Electromagnetic Waves

2.1 MAXWELL'S EQUATIONS

The behavior of the electromagnetic field in any medium is governed by Maxwell's equations. The macroscopic Maxwell's equations in SI units are given by

$$\nabla \cdot \boldsymbol{D}(r,t) = \rho(r,t) \tag{2.1}$$

$$\nabla \cdot \boldsymbol{B}(r,t) = 0 \tag{2.2}$$

$$\nabla \times \boldsymbol{E}(r,t) = -\frac{\delta \boldsymbol{B}(r,t)}{\delta t} \tag{2.3}$$

$$\nabla \times \boldsymbol{H}(r,t) = \frac{\partial \boldsymbol{D}(r,t)}{\partial t} + \boldsymbol{J}(r,t) \tag{2.4}$$

where $\boldsymbol{D}$ denotes the electric displacement, $\boldsymbol{B}$ the magnetic induction, $\boldsymbol{E}$ the electric field, and $\boldsymbol{H}$ the magnetic field. The sources of the field are the free charge density, ρ, and the free current density, $\boldsymbol{J}$.

These are continuous functions in macroscopic electrodynamics. The above equations are written in the space–time domain, but the electromagnetic field and, in particular, its interaction with matter are more easily analyzed in the space–frequency domain. To do this, we use the spectral representation of time-dependent fields, that is, the spectrum $\overline{\boldsymbol{F}}(\boldsymbol{r},\omega)$ of an arbitrary time-dependent field $\boldsymbol{F}(\boldsymbol{r},t)$ that is given by its Fourier transform (see Appendix A*). Conversely, if $\boldsymbol{F}(\boldsymbol{r},\omega)$ is known, the time-dependent field can be calculated by the inverse Fourier transform as shown here:

$$\boldsymbol{F}(r,t) = \int_{-\infty}^{\infty} \overline{\boldsymbol{F}}(r,\omega)e^{-i\omega t}d\omega \tag{2.5}$$

Applying the Fourier transform to Maxwell's equations, one obtains them in the space–frequency domain, and these obviously hold for the spectral components of the electromagnetic field.

$$\nabla \cdot \overline{\boldsymbol{D}}(\boldsymbol{r},\omega) = \overline{\rho}(\boldsymbol{r},\omega) \tag{2.6}$$

---

* Note that in 1-D $F(r,t) \rightarrow F(x,t)$, the Fourier transform of which is $F(k, \omega)$.

15

$$\nabla \cdot \bar{\boldsymbol{B}}(\boldsymbol{r},\omega) = 0 \qquad (2.7)$$

$$\nabla \times \bar{\boldsymbol{E}}(\boldsymbol{r},\omega) = i\omega\bar{\boldsymbol{B}}(\boldsymbol{r},\omega) \qquad (2.8)$$

$$\nabla \times \bar{\boldsymbol{H}}(\boldsymbol{r},\omega) = -i\omega\bar{\boldsymbol{D}}(\boldsymbol{r},\omega) + \bar{\boldsymbol{J}}(\boldsymbol{r},\omega) \qquad (2.9)$$

In the presence of matter, for example, a scattering object or an (in)homogenous medium, an electromagnetic field induces a polarization $\boldsymbol{P}$ and a magnetization $\boldsymbol{M}$ in that object. The vectors $\boldsymbol{D}$ and $\boldsymbol{B}$ take into account this response of the matter, which is driven by the bound and free electron movement in the materials involved. $\boldsymbol{D}$ and $\boldsymbol{B}$ are connected to the polarization and magnetization by the relationships given here.

$$\boldsymbol{D}(\boldsymbol{r},t) = \varepsilon_0\boldsymbol{E}(\boldsymbol{r},t) + \boldsymbol{P}(\boldsymbol{r},t) \qquad (2.10)$$

$$\boldsymbol{B}(\boldsymbol{r},t) = \mu_0[\boldsymbol{H}(\boldsymbol{r},t) + \boldsymbol{M}(\boldsymbol{r},t)] \qquad (2.11)$$

where ε_0 and μ_0 are the electric permittivity and magnetic permeability of in a vacuum, respectively. The relation between $\boldsymbol{E}$ and $\boldsymbol{D}$ and between $\boldsymbol{H}$ and $\boldsymbol{B}$ can be very complicated. For example, the permittivity can be a tensor quantity and so the effective permittivity in one coordinate direction can be quite different in another and we need to understand how an arbitrary incident $\boldsymbol{E}$ field is affected. Worse than this, all materials are to some extent nonlinear, meaning that the permittivity (or permeability) is a function of the field amplitude. Even simple models to describe this become rapidly unwieldy with the permittivity tensor being of rank 3 or 4 depending on whether the field or field intensity drives the changes. (A rank 3 tensor has 27 terms and a rank 4 has 81 terms). In this introductory book, we will ignore nonlinear phenomena and assume that the incident fields are sufficiently weak and that they play no significant role. We do expect, however, that the permittivity or permeability profiles are going to be spatially varying and are probably tensorial in nature.

To begin with, let us consider the electromagnetic field in a simple linear, isotropic, stationary, spatially nondispersive, but temporally dispersive medium. In this case, the polarization and the magnetization are connected to the electric and magnetic field via the following convolution relationships.

$$\boldsymbol{P}(\boldsymbol{r},t) = \varepsilon_0\int_{-\infty}^{t}\chi_e(\boldsymbol{r},t-t')\boldsymbol{E}(\boldsymbol{r},t')\,\mathrm{d}t' \qquad (2.12)$$

$$\boldsymbol{M}(\boldsymbol{r},t) = \int_{-\infty}^{t}\chi_m(\boldsymbol{r},t-t')\boldsymbol{H}(\boldsymbol{r},t')\,\mathrm{d}t' \qquad (2.13)$$

where the response functions χ_e and χ_m, known as the electric and magnetic susceptibilities, vanish for $t' > t$. These equations embody one of the most fundamental principles in physics, namely, causality. The polarization and the magnetization at time t depend on the electric and magnetic fields at all other time instants t before t (origin of temporal dispersion). As can be seen in Appendix A, the consequences of causality are far reaching, leading to a very powerful restriction on the real and imaginary parts of the spectral (i.e., Fourier) representation of these susceptibilities. The spectral susceptibility is an analytic function, and the real and imaginary parts are locked together by an integral transform relationship or dispersion relationship, also known as a Hilbert transform.

The medium may also be spatially dispersive in which case the above relationships would be convolutions over the spatial variable r as well. The effects of spatial dispersion can be most easily understood and observed at interfaces between two distinct media where one might expect some reflection to occur. In media or objects having material fluctuations comparable to the wavelength of the electromagnetic wave being employed, the "reflections" can be quite complex and may well be wavelength-dependent. Once a medium departs from a regular or periodic pattern of material differences, we no longer talk of "diffraction" but of scattering.

Applying the Fourier transform to the expressions for D and B, we can write the constitutive relations in the space–frequency domain as

$$\bar{D}(r,\omega) = \varepsilon_0\varepsilon_r(r,\omega)\bar{E}(r,\omega) \tag{2.14}$$

$$\bar{B}(r,\omega) = \mu_0\mu_r(r,\omega)\bar{H}(r,\omega) \tag{2.15}$$

with $\varepsilon_r = 1 + \chi_e$ and $\mu_r = 1 + \chi_m$ being the relative dielectric permittivity and magnetic permeability of the medium, respectively.

Besides polarization and magnetization, the electromagnetic field may induce currents. The free current density in the material can be divided into two parts: the conduction current density, J_c, induced by an external field and the source current density J_s. The conduction current density is determined by the electric component of the incident field by

$$\bar{J}_c(r,\omega) = \sigma(r,\omega)\bar{E}(r,\omega) \tag{2.16}$$

Starting from Maxwell's curl equations above and using the expressions for D and B, one can derive the inhomogeneous wave equations in the space–time domain for both the electric and magnetic fields. Similarly, using Maxwell's equations and the constitutive relations expressed in the frequency domain, we obtain the wave equations in the space–frequency domain

$$\nabla \times \nabla \times \bar{E}(r,\omega) - k^2(r,\omega)\bar{E}(r,\omega) = i\omega\mu_0\mu_r(r,\omega)\bar{J}_s(r,\omega) \tag{2.17}$$

$$\nabla \times \nabla \times \bar{H}(r,\omega) - k^2(r,\omega)\bar{H}(r,\omega) = \nabla \times \bar{J}_s(r,\omega) \tag{2.18}$$

where $k = k_0 n$ is the wave number in the medium. The term $k_0 = \omega/c_0$ is the wave number in vacuum in which the velocity of the wave is $c_0 = (\varepsilon_0\mu_0)^{-1/2}$, and $n^2 = \varepsilon_r\mu_r$ denotes the square of the refractive index of the material. Making the substitution $\varepsilon_r + i\sigma/\omega_0 \to \varepsilon_r$, we can incorporate the effect of free conducting electrons that might exist in the medium, as described by a very classical model for conduction known as the Drude model.

It is important to stress that these wave equations, which describe the propagation of an electromagnetic field, are valid in an inhomogeneous or scattering media. If we know how the permittivity and permeability (or equivalently the refractive index) vary as a function of space and frequency, we can predict how a wave that is incident in that medium will propagate and scatter. The converse, which is the focus of this book, is a much harder problem.

2.2 GREEN'S FUNCTION

Green's function is very important and gives the electromagnetic field produced by a point source for a given medium or object. In free space or a vacuum, Green's function from a point source is an outgoing spherical wave. Here $\ddot{G}$ is the Green dyadic and $\ddot{I}$ is the unit dyad; a dyadic Green's function is made of three vector Green's functions. Thus, in a homogeneous (infinite) medium, Green's function, $\ddot{G}$, satisfies the equation.

$$\nabla \times \nabla \times \ddot{G}(r,r',\omega) - k^2(\omega)\ddot{G}(r,r',\omega) = \ddot{I}\delta(r - r') \qquad (2.19)$$

where I is the unit tensor and δ denotes the delta function describing the point source located at some position r'. A solution of this equation can be formally written as

$$\ddot{G}(r,r',\omega) = \left[\ddot{I} + \frac{1}{k^2}\nabla\nabla\right]G(r,r',\omega)\mathbf{n}_i \qquad (2.20)$$

where $\mathbf{n}_i$ is a unit vector and where

$$G(r,r',\omega) = \frac{e^{ik|r-r'|}}{4\pi|r - r'|} \qquad (2.21)$$

is the outgoing scalar Green's function that satisfies the wave equation. In terms of Green's function the solution to

$$\nabla \times \nabla \times \bar{E}(r,\omega) - k^2(r,\omega)\bar{E}(r,\omega) = i\omega\mu_0\mu_r(r,\omega)\bar{J}_s(r,\omega) \qquad (2.22)$$

for the current distribution J_s, located in some volume V, becomes

$$\bar{E}(r,\omega) = E_0(r,\omega) + i\omega\mu_0\mu_r(\omega)\int_V \ddot{G}(r,r',\omega)\,J_s(r')\,\mathrm{d}^3r' \qquad (2.23)$$

where r is a point located outside V. The field E_0 is the incident wave, which is typically assumed to be a plane wave, and which one could imagine as being created by a point source at infinite distance away from the medium in volume V.

In electromagnetic propagation and scattering problems, we often assume the source is actually a dipole antenna or emitter. This is a good approximation for sources when not too close to them. Considering a dipole located at r', it has a dipole moment we can denote by q. The corresponding current density is

$$J_s(r) = -i\omega q\delta(r - r') \tag{2.24}$$

The electric field can then be calculated by introducing this expression into Equation 2.23 for E, which gives

$$\bar{E}(r,\omega) = \mu\omega^2\bar{\bar{G}}(r,r_0,\omega)q \tag{2.25}$$

If $E_0 = 0$ (or subtracted out), then Equation 2.25 represents the scattered field.

2.2.1 Solving Differential Equations

Green's function of the wave equation is a solution for the special case when the source term is a delta function in space and/or time.

$$\left[\nabla^2 - \frac{1}{c^2}\frac{\partial^2}{\partial r^2}\right](r,r',t,t') = \delta(r - r')\delta(t - t') \tag{2.26}$$

Green's function is therefore a field radiated by a point source. We can generalize this to arbitrary differential operators. For example, let L_x denote an arbitrary linear differential operators in n variables x (i.e., x_1, x_2, ..., x_n) of order α. In its most general form this would be

$$L_x = \begin{array}{l} a_0 + a_1\dfrac{\partial}{\partial x_1} + a_2\dfrac{\partial}{\partial x_2} + \cdots + a_n\dfrac{\partial}{\partial x_n} + \\[2mm] a_{11}\dfrac{\partial^2}{\partial x_1^2} + a_{12}\dfrac{\partial^2}{\partial x_1\partial x_2} + \cdots + a_{nn}\dfrac{\partial^2}{\partial x_n^2} \\[2mm] +a_{11}\dfrac{\partial^\alpha}{\partial x_1^\alpha} + \cdots + a_{nn}\dfrac{\partial^\alpha}{\partial x_n^\alpha} \end{array} \tag{2.27}$$

Using this operator, we now consider the inhomogeneous differential equation

$$L_x\varphi(x) = -\rho(x) \tag{2.28}$$

where $\rho(x)$ is a source term (as we might expect in Laplace's equation or the inhomogeneous Helmholtz equation). Then we want to find φ accounting for whatever boundary conditions are imposed. We replace $\rho(x)$ by a point source function and then generalize it into distributed source function. This is then the "impulse response" of the system that in optics is a point spread function (PSF)

$$L_xG(x - y) = -\delta(x - y) \tag{2.29}$$

where $G(x,y)$ is Green's function.

Since Green's function behaves like an impulse response function, we assume

$$\varphi(x) = \int G(x, y)\rho(y)\, dy \qquad (2.30)$$

for n integrals where $dy = dy_1, dy_2, \ldots, dy_n$. This can now be substituted back into Equation 2.28 as follows:

$$L_x \varphi(x) = \int L_x G(x, y)\rho(y)\, dy = -\int \delta(x - y)\rho(y)\, dy = -\rho(x) \qquad (2.31)$$

This way of looking at scattering problems fundamentally changes our perspective, whether for a direct or inverse scattering problem. The question now becomes how we find Green's function or the impulse function, which is not so easy in general. We note that if G satisfies the boundary conditions, so does φ.

Let us consider the Fourier transform of G in n-D, that is, n dimensions:

$$G(x, y) = \left(\frac{1}{2\pi}\right)^n \int\int\limits_{-\infty}^{\infty} g(K, y)e^{iK,x}\, dK \qquad (2.32)$$

where $K, x = K_1 x_1 + K_2 x_2 + \cdots$ and $dK = dK_1, dK_2, \ldots$.

Let us recall the following Fourier transform for a delta function as

$$F(\delta(x - y)) = \left(\frac{1}{2\pi}\right)^n \int\int e^{iK(x-y)}\, dK \qquad (2.33)$$

Using this definition in conjunction with Equation 2.29 yields the following relationship:

$$L_x G(x, y) = -\delta(x - y) = \left(\frac{1}{2\pi}\right)^n \int g(k, y)L_x e^{iK,x}\, dK = -\left(\frac{1}{2\pi}\right)^n \int e^{iK,y} e^{iK,x}\, dK \qquad (2.34)$$

Since the operator L_x only operates on x terms, we now have

$$L_x e^{iK,x} = \begin{bmatrix} a_0 + a_1 iK_1 + \cdots + a_n iK_n + \cdots + \\ \cdots + a_{12}(iK_1)(iK_2) + \cdots + a_{nn}(iK_n)^n \cdots + \\ \cdots + a_{n\cdots1}(iK_1)^\alpha + \cdots + a_{n\cdots n}(iK_n)^\alpha \end{bmatrix} e^{iK,x} \qquad (2.35)$$

We can now simplify this and equate to zero which yields

$$\int\limits_{n\ \text{integrals}} \left[g(k, y)[a_0 + a_1 iK_1 + \cdots + a_{nn,n\cdots n}(iK_n)^\alpha] + e^{-iK,y} \right] e^{iK,x}\, dK = 0 \qquad (2.36)$$

This would imply that

$$g(K, y) = \frac{-e^{iK,y}}{a_0 + a_1(iK_1) + \cdots + a_{nn\cdots n}(iK_n)^\alpha} \tag{2.37}$$

Now taking the Fourier transform of Equation 2.37, we have

$$G(x, y) = -\frac{1}{(2\pi)^n} \int \frac{e^{iK,(x-y)} dK}{a_0 + a_1(iK_1) + \cdots + a_{nn\cdots n}(iK_n)^\alpha} \tag{2.38}$$

This is the particular solution. The homogeneous equation can be written as $L_x U(x) = 0$, giving the total solution $G(x,y) + U(x)$ subject to

$$\varphi(x) = U(x) + \int G(x, y)\rho(y)\, dy \tag{2.39}$$

2.2.2 The Integral Equation of Scattering

We can apply a Green's function approach to all wave and particle scattering problems, for example, Schrödinger's equation for a single particle, subject to a potential energy $V(r)$ (e.g., to describe an electron coming into an atom). And for the time-independent case, this reduces to the inhomogeneous Helmholtz equation to

$$\frac{\hbar^2}{2m}\nabla^2\psi(r) + V(r)\psi(r) = E\psi(r) \tag{2.40}$$

where $E = (p^2/2m)$ is the positive energy of the incoming particle and m is its mass in a center of mass coordinate frame. Therefore, we now have

$$(\nabla^2 + K^2)\psi(r) = U(r)\psi(r) = \rho(r) \tag{2.41}$$

$$\psi(r) = \varphi(r) + \int G(r, r')\rho(r')dr' \tag{2.42}$$

Using Fourier transforms and following the earlier analysis, we can write

$$\int g(K', r')(\nabla^2 + K^2)e^{iK',r} dK' = -\int e^{-iK',r'}e^{iK,r}dK' \tag{2.43}$$

$$\nabla^2 e^{iK',r} = -(K')e^{iK',r} \tag{2.44}$$

$$\int g(K', r')(K^2 - (K')^2)e^{iK',r}dK' = -\int e^{-iK',r}e^{iK,r}dK' \tag{2.45}$$

$$g(K', r') = \frac{-e^{-iK',r'}}{K^2 - (K')^2} \tag{2.46}$$

$$G(r,r') = -\left(\frac{1}{2\pi}\right)^3 \int \frac{e^{-iK',(r-r')}dK'}{K^2 - (K')^2} \qquad (2.47)$$

$$\psi(r) = \varphi(r) - \int G(r,r')U(r')\psi(r')dr' \qquad (2.48)$$

where $\psi(r)$ is the total wave, $\varphi(r)$ is the incident or unscattered wave and the integral represents the total scattered wave. If the scattering is weak then $\psi \simeq \varphi$, which is the first Born approximation. Examples of when the first Born approximation might be valid include the case when an electron incident on an atom has extremely high energy or when an electron approaches a nucleus and needs to be very close for a significant interaction to take place. In electromagnetic interactions, x-ray illumination of materials is typically a weak interaction.

Now we consider

$$G(r,0) = -\frac{1}{(2\pi)^3} \int \frac{e^{iK',r}dK'}{K^2 - (K')^2}$$

$$= -\frac{1}{(2\pi)^3} \int_{K'=0,\theta=0,\varphi=0}^{\infty} \int_{}^{\pi} \int_{}^{2\pi} \frac{e^{iK'r\cos\theta}(K')^2 \sin\theta\, d\theta\, dK'\, d\varphi}{K^2 - (K')^2} \qquad (2.49)$$

Here the expression is written in spherical polar coordinates with $dK' = (K')^2 \sin\theta\, d\theta\, dK'\, d\varphi$.

If there is no φ dependence (i.e., only the scattering angle important and not direction) then

$$\int_0^{\pi} e^{iK'r\cos\theta}\sin\theta\, d\theta = \frac{1}{iK'r}(e^{iK'r} - e^{-iK'r}) \quad \text{if} \quad \int e^{iK'x}dx = -\int_0^{\pi} e^{iK'r\cos\theta}d(\cos\theta) \qquad (2.50)$$

$$G(r,0) = \frac{-2\pi}{(2\pi)^3} \int_0^{\infty} \frac{e^{iK'r} - e^{-iK'r}K'dK'}{K^2 - (K')^2}$$

$$= \frac{1}{4\pi^2 i}\frac{1}{r} \int_0^{\infty} \frac{[e^{iK'r} - e^{-iK'r}]K'dK'}{K^2 - (K')^2} \qquad (2.51)$$

This function has poles at $K' = \pm K$. The limit can now be taken as follows:

$$G(r,0) = -\lim_{\varepsilon \to 0} \frac{1}{4\pi i}\frac{1}{r} \int_0^{\infty} \frac{(e^{iK'r} - e^{-iK'r})K'dK'}{(K^2 + i\varepsilon) - (K')^2} \qquad (2.52)$$

There are three different approaches to taking this limit, giving three different answers. We could write $(K^2 - i\varepsilon) - (K')^2$ which switches from outgoing to ingoing wave, that is, this switches the time origin.

It could now be written that

$$(K^2 - i\varepsilon) - (K')^2 = \left[((K^2 - i\varepsilon)^{1/2} - K')((K^2 - i\varepsilon)^{1/2} + K')\right]$$

$$\simeq \left[\left(K + \frac{i\varepsilon}{2K}\right) - K'\right]\left[\left(K + \frac{i\varepsilon}{2K}\right) + K'\right] \qquad (2.53)$$

which has poles at $K' = \pm(K + (i\varepsilon/2K))$ giving

$$G(r,0) = -\lim_{\varepsilon \to 0}\frac{1}{2}\int_{-\infty}^{\infty}\frac{1}{4\pi^2 ir}\frac{(e^{iK'r} - e^{-iK'r})K'dK'}{\left[\left(K + \dfrac{i\varepsilon}{2K'}\right) + K'\right]\left[\left(K + \dfrac{i\varepsilon}{2K'}\right) - K'\right]} \qquad (2.54)$$

and then use Cauchy's residue theorem

$$\left[\oint_C f(z)dz = 2\pi i\sum \text{Residues}\right]$$

in the upper-half plane (uhp) of the complex plane and the

$$\left\{\text{Residue} = a_{-1} = \frac{1}{(m-1)!}\lim_{z \to a}\left(\frac{d}{dz}\right)^{m-1}(z-a)^m f(z)\right\}$$

Hence, the

$$\text{residue} = \frac{1}{0!}\lim_{K' \to K + \frac{i\varepsilon}{2K}}\left[K' - K + \left(\frac{i\varepsilon}{2K}\right)\right]f(z)$$

Assuming Jordan's lemma holds, namely, that as $R \to \infty$, $\int_{uhp} = 0$. Therefore, residue = $R_1 + R_2$ where

$$R_1 = \lim_{K' \to K + \frac{i\varepsilon}{2k}}\frac{-e^{iK'r}K'}{K + (i\varepsilon/2K) + K'}$$

$$R_1 = -\frac{e^{i(K+(i\varepsilon/2K))r}(K + (i\varepsilon/2K))}{2K + (i\varepsilon/K)} \qquad (2.55)$$

and as $\varepsilon \to 0, R_1 \to -(1/2)e^{iKr}$.

The second integral is $-\int_{-\infty}^{\infty}\dfrac{e^{-iK'r}K'dK'}{[K + (i\varepsilon/2K) + K'][K + (i\varepsilon/2K) - K']}$

If we let $K' = -K'$, then this integral has already been evaluated and so, $R_1 = R_2$

$$G(r,0) = -\frac{1}{2}\frac{1}{4\pi^2 i}\frac{1}{r}\, 2\pi i(R_1 + R_2) = \frac{1}{4\pi}\frac{e^{iKr}}{r} \qquad (2.56)$$

In general:

$$G(r,r') \rightarrow G^+(r,r') = \frac{1}{4\pi} \frac{e^{iK|r-r'|}}{|r-r'|} \qquad (2.57)$$

We note that an equation of the form $\nabla^2\psi - (1/v^2)(\partial^2\psi/\partial t^2) = 0$ (where $\nabla^2\psi = (\partial^2\psi/\partial x^2)$) is solved by setting $x - vt = p$ and $x + vt = q$; therefore $\partial^2\psi/\partial p\partial q = 0$. Having the solution $\psi = \psi_1(p) + \psi_2(q) = \psi_1(x - vt) + \psi_2(x + vt) = \psi^+ + \psi^-$, we see that we have, as expected, an ingoing and an outgoing wave. From this it can be written

$$\psi^\pm(r) = \varphi(r) - \int \frac{e^{\pm iK|r-r'|}}{|r-r'|} U(r')\psi^\pm(r')d^3r' \qquad (2.58)$$

and this is an inhomogeneous Fredholm type of equation of the second kind. If $U(r')\psi^+(r')$ decreases rapidly as $|r'| \rightarrow \infty$ (e.g., if it vanishes for $|r'| > R$), then letting $|r| \rightarrow \infty$ (which is equivalent to $|r'| \rightarrow 0$) leads to

$$\psi^+(r) \rightarrow \varphi(r) - \frac{1}{4\pi} \frac{e^{iKr}}{r} \underbrace{\int U(r')\psi^+(r')d^3r'}_{\text{constant}}$$

$$\rightarrow \varphi(r) + \frac{1}{4\pi}C' \frac{e^{iKr}}{r} \qquad (2.59)$$

$$\rightarrow \varphi(r) + C \frac{e^{iKr}}{r}$$

where C is the scattering amplitude.

Similarly, it follows that

$$\psi^-(r) \rightarrow \varphi(r) + C \frac{e^{-iKr}}{r} \qquad (2.60)$$

At $|r| \rightarrow \infty$, by looking at the time dependence and locus of the phase, it can be determined whether we have an ingoing or outgoing wave; for example, with Schrödinger's equation,

$$\psi(r,t) = \psi(r)e^{-i(E/\hbar)t} = \psi(r)e^{-i\omega t}$$

Therefore

$$\psi^+(r,t) = e^{i(Kr-\omega t)} + \frac{Ce^{i(Kr-\omega t)}}{r} \qquad (2.61)$$

For ψ^+ one can see that Kr increases as ωt increases, corresponding to an outgoing wave (i.e., its constant phase front moves out). This would typically be written as

$$\psi^+ = \psi(r) = e^{iKz} + \frac{e^{iKr}}{r} f(\theta,\varphi) \qquad (2.62)$$

We can now define the differential cross section as $d\sigma(\theta,\varphi)/d\Omega \equiv |f(\theta,\varphi)|^2$. This cross section is defined as $\sigma = $ (power scattered/power incident). We also note that

$$\sigma = \int d\sigma = \int |f(\theta,\varphi)|^2 \, d\Omega = \int f^*(\theta,\varphi)f(\theta,\varphi)d\Omega = \frac{4\pi}{K}\operatorname{Im}f(0,0) \qquad (2.63)$$

that is, the imaginary part of the forward scattering amplitude measures the loss of intensity that the incident beam suffers due to scattering. This is known as the Optical Theorem.

In the scattering examples that follow, we usually also make the so-called far-field approximation. We assume an incident plane wave of the form e^{iKz} as

$$\psi(r) = e^{iKz} - \frac{1}{4\pi}\int \frac{e^{iK|r-r'|}}{|r-r'|}U(r')\psi(r')d^3r' \qquad (2.64)$$

and we assume that the "potential" $U(r)$ has limited range such that the integral is over some finite range of the variable r'. From this assumption, we can write

$$|r - r'| = |\underline{r} - \underline{r}'| = \left[|r|^2 + |r'|^2 - 2r - r'\right]^{1/2}$$

$$= |r|\left[1 - \frac{2\underline{r} - \underline{r}'}{|r|^2}\right]^{1/2}$$

$$= |r|\left[1 - \frac{\underline{r} - \underline{r}'}{|r|^2} + \cdots\right]$$

$$= |r| - (\hat{\underline{r}} - \hat{\underline{r}}')|r'|$$

Therefore

$$\frac{e^{iK|\underline{r}-\underline{r}'|}}{|\underline{r}-\underline{r}'|} \rightarrow \frac{e^{iKr}e^{-iKr'(\hat{\underline{r}}-\hat{\underline{r}}')}}{r} = \frac{e^{iKr}e^{-iK'r'}}{r} \qquad (2.65)$$

This now allows us to write

$$\underbrace{\psi(r)}_{r\to\infty} \rightarrow e^{iKz} - \frac{e^{iKr}}{4\pi r}\int e^{-iK'r'}U(r')\psi(r')d^3r' \qquad (2.66)$$

so that now the scattering amplitude becomes

$$f(\theta) = -\frac{1}{4\pi}\int e^{-iK'r'}U(r')\psi(r')d^3r' \qquad (2.67)$$

This is simply the Fourier transform of the product $U\psi$ and is an exact representation in the far field, also referred to as the Fraunhofer approximation. $U\psi$ is referred to as the object wave or as a secondary source since it is U

that is the cause of the scattering and that is the function we wish to recover, whereas our measurements are of the product $U\psi$ and we do not know ψ. If ψ is known, for example, whether it is a constant or a plane wave, then we can invert this for the scattering function.

2.3 PLANE WAVES

Let us now consider a quasi-monochromatic wave. A truly monochromatic wave is not physical since it would be noncausal. However, it is very convenient to consider an extremely narrowband source that approximately oscillates at a single constant frequency. Our knowledge of Fourier analysis tells us that a finite bandwidth source can always be represented by a weighted sum of single frequency waves. This greatly simplifies our analysis and facilitates our intuitive understanding of scattering mechanisms and inverse scattering procedures. We have already neglected nonlinear processes in our media and so we have a linear system, but we do still allow for temporal dispersion, that is, that different frequencies might see different refractive index values.

A quasi-monochromatic field component at a frequency ω will have the form

$$F(r,t) = \bar{F}(r,\omega)e^{i\omega t} \tag{2.68}$$

where the spatial part $F(r, \omega)$ is complex, that is, it has a real and imaginary part or, equivalently, a magnitude and a phase. At high frequencies, for example, greater than 1 THz, we tend to measure the averaged energy or time-averaged Poynting vector associated with the electromagnetic field, $E \times H$, which in homogeneous measurement space is proportional to $|E|^2$. Consequently, when we calculate this value, we lose the very important information about the complex field, namely, its phase. For time-harmonic fields, the time-dependent Maxwell's equations take the form

$$\nabla \cdot D(r,\omega) = \rho(r,\omega) \tag{2.69}$$

$$\nabla \cdot B(r,\omega) = 0 \tag{2.70}$$

$$\nabla \times E(r,\omega) = i\omega B(r,\omega) \tag{2.71}$$

$$\nabla \times H(r,\omega) = -i\omega D(r,\omega) + J(r,\omega) \tag{2.72}$$

These equations are equivalent to those of the spectral components of an arbitrary time-dependent field shown above. Consequently, the solutions for $E(r, \omega)$ and $H(r, \omega)$ are identical to the solutions for $\bar{E}(r,\omega)$ and $\bar{H}(r,\omega)$.

For dielectric materials the free charge density ρ and the source current density J_s are zero. This is valid also for materials having appreciable conductivity, that is, for metals. If the medium is homogenous, that is, if the material parameters do not depend on r, the wave equations reduce to the so-called homogenous Helmholtz equations.

$$[\nabla^2 + k^2(\omega)]\boldsymbol{E}(\boldsymbol{r},\omega) = 0 \tag{2.73}$$

$$[\nabla^2 + k^2(\omega)]\boldsymbol{H}(\boldsymbol{r},\omega) = 0 \tag{2.74}$$

The most useful solution for these Helmholtz equations is the quasi-monochromatic plane wave and the spatial part of the solutions is

$$\boldsymbol{E}(\boldsymbol{r},\omega) = \boldsymbol{E}_0(\omega)e^{i\boldsymbol{k}\cdot\boldsymbol{r}} \tag{2.75}$$

$$\boldsymbol{H}(\boldsymbol{r},\omega) = \boldsymbol{H}_0(\omega)e^{i\boldsymbol{k}\cdot\boldsymbol{r}} \tag{2.76}$$

where $\boldsymbol{E}_0(\omega)$ and $\boldsymbol{H}_0(\omega)$ are the complex electric and magnetic field amplitudes, that is, they are complex, having a magnitude and phase. The vector $\boldsymbol{k}$ is the wave vector which must satisfy

$$\boldsymbol{k}\cdot\boldsymbol{k} = k_0^2\varepsilon_r(\omega)\mu_r(\omega) \tag{2.77}$$

This expression is very important since it effectively describes all the propagation characteristics of the medium and is known as the dispersion relationship, indicating the necessary relationship that must exist between the frequency and the wave number. Just to confuse matters, the relation between the real and imaginary parts of the fields, as a function of frequency, also satisfy a dispersion relation (the Hilbert transform mentioned earlier), and if the medium or scatterer is discrete, that is, is embedded in a homogeneous background, then one can show that the real and imaginary parts of the field as a function of $\boldsymbol{k}$ will also satisfy a similar Hilbert transform relation. We will revisit this kind of dispersion relationship later in the book, since it provides a key to not only recovering unmeasured phase information from measured intensity data, but also toa mechanism for solving the inverse scattering problem.

Finally, we note the concept of the angular plane wave representation of a scattered or propagating field. By decomposing an arbitrary field into a superposition of waves in space, just as one might in frequency, we gain a powerful tool and additional insight. The angular spectrum representation is a particularly useful technique for understanding the propagation of optical fields in homogenous media. It is the expansion of an electromagnetic field in a weighted summation of plane waves with variable amplitudes and propagation directions.

If we consider a quasi-monochromatic field in a region $0 \le z \le Z$ with sources located outside of this region, then the spatial part of the field satisfies the Helmholtz equations. We can express the field by a Fourier integral

$$E(x,y,z,\omega) = \int\limits_{-\infty}^{\infty}\!\!\int E(k_x,k_y;z,\omega)e^{i(k_x x + k_y y)}\mathrm{d}k_x\mathrm{d}k_y \tag{2.78}$$

Here $\boldsymbol{k}_x$ and $\boldsymbol{k}_y$ are spatial frequencies corresponding to the real space coordinates x and y. Spatial frequencies describe some number of spatial periods per unit distance, just as a temporal frequency describes some number of temporal cycles or periods of a wave per unit time. If we insert this expression into the Helmholtz equation, we obtain a differential equation for $E(k_x, k_y; z, \omega)$ which is

$$\frac{\partial^2 E(k_x, k_y; z, \omega)}{\partial^2 z} + k_z^2 E(k_x, k_y; z, \omega) = 0 \qquad (2.79)$$

where

$$k_z = +[k^2 - (k_x^2 + k_y^2)]^{1/2} \quad \text{when } k_x^2 + k_y^2 \leq k^2 \qquad (2.80)$$

$$k_z = +i[(k_x^2 + k_y^2) - k^2]^{1/2} \quad \text{when } k_x^2 + k_y^2 > k^2 \qquad (2.81)$$

and $k = k_0 n$ is the wave number in the $0 < z < Z$ region. Using the general solution for this differential equation, the field in this region can be expressed as

$$E(x, y, z, \omega) = \int\int_{-\infty}^{\infty} A(k_x, k_y; \omega) e^{i(k_x x + k_y y + k_z z)} dk_x dk_y$$

$$+ \int\int_{-\infty}^{\infty} B(k_x, k_y; \omega) e^{i(k_x x + k_y y - k_z z)} dk_x dk_y \qquad (2.82)$$

where $A(k_x, k_y; \omega)$ and $B(k_x, k_y; \omega)$ are arbitrary functions. This expression is known as the angular spectrum representation of the E field. When the refractive index is real and positive, the z-component of the wave vector, k_z, is either real or purely imaginary. Therefore, this expression for the fields represents that wave field in terms of four types of plane wave solutions:

1.

$$e^{i(k_x x + k_y y)} e^{ik_z z} \quad \text{where } k_z = +\left[k^2 - \left(k_x^2 + k_y^2\right)\right]^{1/2} \text{ and } k_x^2 + k_y^2 \leq k^2 \quad (2.83)$$

   These solutions are homogenous plane waves that propagate from the boundary plane $z = 0$ toward the boundary plane $z = Z > 0$.

2.

$$e^{i(k_x x + k_y y)} e^{ik_z z} \quad \text{where } k_z = +i\left[(k_x^2 + k_y^2) - k^2\right]^{1/2} \text{ and } k_x^2 + k_y^2 > k^2 \quad (2.84)$$

   These solutions are evanescent waves that decay exponentially from plane $z = 0$ toward the boundary plane $z = Z > 0$.

3.

$$e^{i(k_x x + k_y y)} e^{ik_z z} \quad \text{where } k_z = +\left[k^2 - \left(k_x^2 + k_y^2\right)\right]^{1/2} \text{ and } k_x^2 + k_y^2 \leq k^2 \quad (2.85)$$

These are plane waves that propagate from $z = Z > 0$ toward $z = 0$.

4.

$$e^{i(k_x x + k_y y)} e^{ik_z z} \quad \text{where } k_z = +\left[k^2 - \left(k_x^2 + k_y^2\right)\right]^{1/2} \text{ and } k_x^2 + k_y^2 \le k^2 \quad (2.86)$$

These are evanescent waves decaying exponentially from the plane $z = Z$ toward the plane $z = 0$.

In an absorbing material, the permittivity and the permeability are complex quantities. Consequently, the wave vector becomes complex and the waves decay. The propagating and evanescent waves can be considered slow-decaying and fast-decaying waves.

The angular spectrum representation for a field that propagates into the half-space $z \ge 0$ and whose sources are located in $z < 0$ is obtained by neglecting the latter term. The spectral amplitudes $A(k_x, k_y; \omega)$ of each plane wave component are given by the Fourier transform of the field in the plane $z = 0$. Thus, if the field is known in the plane $z = 0$, it is known throughout the half space $z > 0$ by using this angular spectrum representation.

We also note that the boundary conditions for an electromagnetic field at the interface between two media are derived from the integral forms of Maxwell's equations. The boundary conditions are valid for both the time-dependent fields and their spectral components, and they take the form

$$\hat{n} \times (\boldsymbol{E}_2 - \boldsymbol{E}_1) = 0 \qquad\qquad (2.87)$$

$$\hat{n} \times (\boldsymbol{H}_2 - \boldsymbol{H}_1) = \boldsymbol{J}_{su} \qquad\qquad (2.88)$$

$$\hat{n} \cdot (\boldsymbol{D}_2 - \boldsymbol{D}_1) = \rho_{su} \qquad\qquad (2.89)$$

$$\hat{n} \cdot (\boldsymbol{B}_2 - \boldsymbol{B}_1) = 0 \qquad\qquad (2.90)$$

where $\hat{n}$ is the unit vector normal to the interface pointing from the input medium and into a second medium. The vector $\boldsymbol{J}_{su}$ denotes the surface current density, and ρ_{su} is the surface charge density. The fields are governed by Maxwell's equations, which is why the boundary conditions are not independent of each other.

2.4 EVANESCENT AND PROPAGATING WAVES

In this chapter, we ignore scattering objects that vary in time and make the assumption that illumination is by using a plane (quasi-) monochromatic wave. This greatly simplifies the theoretical model. Assuming linearity, pulsed illumination could be modeled using a set of waves with different frequencies and an incident nonplane wave can similarly be decomposed into a set of weighted angular plane waves. The scattering from an object can be assumed to generate an infinite set of plane waves not all of which are propagating. This can be expressed mathematically as

$$E(r) = \sum_{k_x,k_y} E_0(k_x,k_y)e^{i(k_x x + k_y y + k_z z - \omega t)}$$

(2.91)

where

$$k_z = \sqrt{\omega^2/c_0^2 - k_x^2 - k_y^2}$$

(2.92)

However, k_z takes only real values if

$$\frac{\omega^2}{c_0^2} = \frac{2\pi}{\lambda} > k_x^2 + k_y^2$$

(2.93)

Those waves or spatial frequencies with imaginary propagation constants carry the highest resolution information about the scattering object, but decay exponentially away from the object. Near-field measurements, that is, gathering data within a distance of ~λ from the surface of the scatterer, can capture some of this information. More typically, the scattered field is measured many wavelengths from the scatterer, and only the propagating scattered waves contribute to the signal. This limits the resolution available about the scattering target to ~$\lambda/2$. As will be discussed in later chapters, many methods to improve the resolution of an image exist and most implicitly involve using some kind of cost function to extract a single solution from a space of solutions which minimizes (or maximizes) some meaningful quantity such as the total energy or entropy of a data consistent (or almost consistent) image. The incorporation of prior knowledge (when available) about the anticipated shape or structure of the scattering object can greatly help here, but this is frequently only known in special cases such as medical or nondestructive testing applications of inverse scattering.

Propagation from the near to the far field is easily done by using a plane wave expansion of the propagating wave and recognizing that propagation in a homogeneous medium is a convolution which can be executed by taking the Fourier transform of the wave in one plane, applying a phase shift to each plane wave component corresponding to the distance the wave is to be propagated, and then inverse transforming the results (see Appendix C).

Three

Scattering Fundamentals

3.1 MATERIAL PROPERTIES AND MODELING

3.1.1 The Model for Conductivity

The most basic equation of motion is that given by Newton, and it describes how a charged particle is affected by an electric field:

$$m\left(\frac{dv}{dt} + \frac{v}{\tau}\right) = eE \tag{3.1}$$

where m is the electron mass, e its charge, v the velocity and t is time; here we add a damping term with relaxation time τ. Throughout this book we will assume waves to have a temporal variation, $e^{j\omega t}$ (ω is the angular frequency). From the above equation, we can now write

$$v = \frac{e}{m}\frac{E}{j\omega + (1/\tau)} \tag{3.2}$$

and the current density is expressed as

$$J = Nev = \frac{Ne^2\tau}{m}\frac{E}{1 + j\omega\tau} \tag{3.3}$$

which, as we have already seen, can be written as

$$J = \sigma E \quad \text{where } \sigma = \frac{\sigma_0}{1 + j\omega\tau} \quad \text{and} \quad \sigma_0 = \frac{Ne^2\tau}{m}\text{(electrical conductivity)} \tag{3.4}$$

3.1.2 Time-Dependent Maxwell's Equations

Let us list the time-dependent forms for Maxwell's equations:

$$\text{Electric flux density (units C/m}^2)\quad \mathbf{D} \rightarrow \nabla \cdot \mathbf{D}(t) = \rho_v(t) \tag{3.5}$$

$$\text{Electric field intensity (units V/m)}\quad \mathbf{E} \rightarrow \nabla \cdot \mathbf{B}(t) = 0 \tag{3.6}$$

$$\text{Magnetic flux density (units W/m}^2)\quad \mathbf{B} \rightarrow \nabla \times \mathbf{E}(t) = -\frac{\partial \mathbf{B}(t)}{\partial(t)} \tag{3.7}$$

$$\text{Magnetic field intensity (units A/m)}\quad \mathbf{H} \rightarrow \nabla \times \mathbf{H}(t) = \mathbf{J}(t) + \frac{\partial \mathbf{D}(t)}{\partial(t)} \tag{3.8}$$

It is important to recall that ρ_v; the electric charge density (C/m^2) and J the electric current density (A/m^2) are sources that can induce electromagnetic fields or be induced by the fields.

Fundamentally, we can say that materials are composed of charged particles, which are displaced by the applied fields, which in turn modify the propagation of these fields. To describe this at the microscale (unit: atom, molecule, etc.), we need to define the polarizabilities, or equivalently, a scattering parameter, such as a scattering cross section, which will be introduced later. At the macroscale (i.e., bulk material), we introduce the constitutive parameters we see above:

   ε: dielectric permittivity, a measure of how well a material can store energy imposed by an electric field; it is indirectly associated with capacitance.
   μ: permeability, a measure of how efficiently energy is stored from an applied magnetic field; it is indirectly associated with inductance.
   σ: conductivity, the quantity arising from free charges that lead to currents in the presence of an applied field.

where

$$D(t) = \varepsilon(t) * E(t) \tag{3.9}$$

$$B(t) = \mu(t) * H(t) \tag{3.10}$$

$$J(t) = \sigma(t) * E(t) \tag{3.11}$$

3.1.3 Effective Permittivity and Conducting Medium

Using the above identities, we can write

$$J + j\omega\varepsilon E \equiv j\omega\varepsilon_{\text{eff}} E \tag{3.12}$$

where

$$\varepsilon E = \varepsilon + \frac{\sigma_0}{j\omega}\frac{1}{1 + j\omega\tau} \quad \text{since } J = \sigma E \tag{3.13}$$

Likewise

$$\sigma = \frac{\sigma_0}{1 + j\omega\tau} \quad \text{and} \quad \sigma_0 = \frac{Ne^2\tau}{m} \tag{3.14}$$

The following simplifications are now evident:

1. At low frequencies this simplifies to

$$\varepsilon_{\text{eff}} = \varepsilon - \left(\frac{j\sigma_0}{\omega}\right) \tag{3.15}$$

2. At high frequencies this simplifies to

$$\varepsilon_{\text{eff}} = \varepsilon_0 \left(1 - \frac{\omega_p^2}{\omega^2} \right) \tag{3.16}$$

We can now use this expression to define

$$\omega_p^2 = \frac{Ne^2}{\varepsilon_0 m} \tag{3.17}$$

where ω_p is defined as the plasma frequency. It should now be noted that when $\omega < \omega_p$, then $\varepsilon_{\text{eff}} < 0$ and when $\omega > \omega_p$, then $\varepsilon_{\text{eff}} > 0$.

It will be shown that when we develop a wave equation, that wave equation still holds when we replace ε by this new quantity ε_{eff}. However, when ε_{eff} is complex, then the wavenumber or propagation constant k is complex. If k has a nonzero imaginary part, then there is no propagation and the wave is said to be evanescent, which means it exponentially decays in the direction of propagation. We will see later in the context of imaging, that k becomes imaginary when very high resolution information (i.e., high "spatial" frequency information) is imposed on an applied propagating field.

3.1.4 Increasing N and Local Fields

If polarizable regions in a material do not interact, as might be the case in a gas, then we can express the macroscopic properties of a bulk material or object we are illuminating in terms of the polarizability of each unit, for example, atom or (nonpolar) molecule (Table 3.1). We expressed that $\bar{P} = N\alpha\bar{E}$ where α was the polarizability of the "unit" in the material and N, the number density of these polarizable units. From this we can write $\varepsilon_r = 1 + N\alpha/\varepsilon_0$ and since $\varepsilon_r = n^2$, where n is the refractive index, it follows that

$$n = [1 + N\alpha/\varepsilon_r]^{1/2} \sim 1 + N\alpha/2\varepsilon_0 \tag{3.18}$$

This last step is from truncating a Taylor series expansion, which is justified for a "low concentration" of scattering units (such as a gas) and defines it as a weak scatterer showing that the index is simply proportional to N.

As the density of the units increases, we cannot neglect the effects of the local field, coupling, and multiple scattering between neighboring units or inhomogenieties. We should now write $\bar{P} = N\alpha E_{\text{loc}} = N\alpha g E_{\text{ext}}$ and $\varepsilon_r = 1 + N\alpha g/\varepsilon_0$. A very simple model to approximate these local field effects is to define a

Table 3.1 Sample Material Properties

	Form	Density	N_x/ε_0	ε_r
CS_2	Gas	0.00339	0.0029	1.0029
	Liquid	1.293	1.11	2.76
O_2	Gas	0.00143	0.000523	1.000523
	Liquid	1.19	0.435	1.509

spherical region inside the medium, which is large compared to any one unit. The field inside a uniformly polarized sphere can be approximated by a dipole given by $4\pi\alpha^3\bar{P}/3$, and after some integrations and manipulations, one can calculate the local field at a unit in the center of this sphere. The local field, E_{loc}, is the macroscopic field E_{ext}, minus the contribution due to the units inside the sphere. If the units are spherically symmetric (which might well not be the case!), then the surrounding units act in a spherically symmetric fashion and we find $g = (2 + \varepsilon_r)/3$ and

$$E_{loc} = E_{ext} + \frac{P}{3\varepsilon_0} \tag{3.19}$$

Strictly speaking, the polarization is proportional to the local electric field, so

$$P = N\alpha_e E_{loc} \tag{3.20}$$

where α_e now represents an atomic or molecular polarizability. This now allows us to write

$$P = E_{ext} \frac{N\alpha_e}{1 - (N\alpha_e/3\varepsilon_0)} \tag{3.21}$$

From the definition of ε_{eff}, we can now write that

$$D = \varepsilon_{eff} E_{ext} = \varepsilon_0 E_{ext} + P \tag{3.22}$$

and

$$\varepsilon_{eff} = \varepsilon_0 + \frac{N\alpha_e}{1 - (N\alpha_e/3\varepsilon_0)} \tag{3.23}$$

Thus, the relative permittivity is given by

$$\varepsilon_r = \frac{\varepsilon_{eff}}{\varepsilon_0} = 1 + \frac{N\alpha_e/\varepsilon_0}{1 - (N\alpha_e/3\varepsilon_0)} \tag{3.24}$$

which is the well-known Clausius–Mossotti relation.

If we now solve for $N\alpha_e$, we have

$$N\alpha_e = 3\varepsilon_0 \frac{\varepsilon_{eff} - \varepsilon_0}{\varepsilon_{eff} + 2\varepsilon_0} = 3\varepsilon_0 \frac{\varepsilon_r - 1}{\varepsilon_r + 2} \tag{3.25}$$

Knowing N and α_e we can calculate the relative permittivity when the various approximations made are satisfied. Notice that when $(N\alpha_e/\varepsilon_0) < 3$, the relative permittivity is negative. For naturally occurring materials, we can replace the number density N by the bulk density multiplied by Avogadro's number and divided by the molar mass.

3.2 WEAK SCATTERERS

Note that for all dielectric media for which $\varepsilon_r > 1$ it is apparent that the local field is greater than the external applied field. This sometimes causes confusion because we usually think of the dielectric constant of a material as providing a screening of the applied field, thereby making the local field smaller. However, this ignores the polarization of the material itself, and inside the sphere, we observe the external field plus the field due to the medium's polarization. It will be explained later how, in the frequency domain, the real and imaginary parts of the complex permittivity are locked together through dispersion relations (Kramer–Kronig dispersion relations or in the mathematics community, Hilbert transforms). This is very fundamental, being based on the causality principle, and very important since it allows one to compute the real part from the imaginary part and vice versa. It will be shown that controlling spectral absorption profiles directly determines spectral refractive index profiles, and we usually are trying hard to manage the absorption and dispersion in the materials. Another remark worth noting is that while we have used the polarizability α_e here, it could represent a polarizability arising from a number of different physical mechanisms including electronic, ionic, permanent dipole rotational, or displacement effects.

We will assume for the sake of simplicity that our scattering objects will not have a permanent dipole moment. There are models to deal with this case and incorporate their thermally randomized alignment known as the Debye model, but even this fails for simple units like H_2O because of the complexity of local interactions. The Onsager model attempts to create a cavity around a single unit, and the Kirkwood model creates a small cluster around the unit. Both may be important to help advance models for strongly resonant atoms or structures. It is very interesting to note also that in practice when it comes to coupling effects between neighboring units, usually less than 10 neighbors are important and sometimes only immediate nearest neighbors are important.

3.3 SCATTERING FROM COMPACT STRUCTURES

Most objects to be imaged by electromagnetic waves are 3-D in nature, and there is to date no well-developed diffraction or scattering theory for such objects. Solutions of the scattering problem based on Maxwell's equations have been obtained for some idealized objects such as spheres or cylinders (Barber and Yeh, 1975; Logan, 1965). However, even in these cases, the solutions are generally expressed as infinite series of special functions, which makes the interpretation of the solutions difficult. As a result of this, many approximate methods have been proposed which do allow the scattered field to be expressed analytically or which are convenient for numerical experiments (Keller, 1961).

Inverse scattering methods for studying the detailed local structure inside penetrable compact objects are the focus of much of this text. There are two approaches to determining the internal structure of such objects from scattered and transmitted field data. The first relies on a geometrical optics description, which requires the assumption that the wavelength used is considerably smaller than the structural detail of interest. This assumption is adequate at x-ray wavelengths when used in computerized tomography mentioned previously. When the wavelength is not relatively small, scattering and diffraction effects become important. It is this specific case that we will focus on here.

The problem of determining the image of an object from scattered field measurements has received a lot of attention lately because the associated applications are so vast. For objects that are sufficiently weakly scattering, it is well known that the inverse scattering problem can be linearized and thus becomes readily soluble using the first Born or Rytov approximations (see later). Object reconstruction can be achieved by extracting the Fourier data of the object from the scattered field. This is available over semicircular arcs in the Fourier plane (k-space) of the object, which normally often requires interpolation and extrapolation to improve the image estimate. The more the measurements of the scattered field that are made, the more the points obtained in k-space. For compact objects, the Fourier transform is analytic, or more precisely, an entire function of exponential type. It is this fundamental underlying property that leads to the definition of sampling theorems and provides effective methods of extrapolation and interpolation. In principle, knowing an analytic function over a small but continuous interval of k-space should allow it to be extrapolated by analytic continuation throughout k-space. Of course, the fact that the data are measured at discrete points and are finite in number means that we have an infinite number of possible solutions. The analyticity does provide an overall constraint and the compact support, even if it is only approximately known, provides considerable prior knowledge that can be exploited. A support constraint not only allows an (in principle) adequate sampling rate to be defined, but also allows the Fourier basis being used to be modified to a more appropriate and more effective basis to be defined. We discuss these concepts more in Chapter 4. Improved sampling strategies combined with more effective representations for the object suggest that we might be able to get better images with fewer measurements and there is some truth to this. The most recent implementation of these ideas is described and explored in Chapters 6 and 10.

3.3.1 Scattering from Obstacles

When a field excites a dipole, energy is transferred and seen as a loss or "extinction" of the incident wave. Time-averaged scattered power is defined as $P_{scat} \equiv \sigma_{scat} I_0$ with some 3-D scattered field pattern. There may also be some absorption $P_{abs} \equiv \sigma_{abs} I_0$ and a total extinction cross section given by $P_{ext} \equiv \sigma_{ext} I_0$ where $\sigma_{ext} = \sigma_{scat} + \sigma_{abs}$. The differential cross section is defined in Figure 3.1.

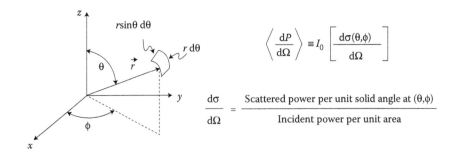

Figure 3.1   Illustration of differential cross section.

The scattering cross section is defined as $\sigma = N_s/N_i$ where N_s is the number of scattered photons/area and N_i the number of incident photons per unit area. There are several different classifications of scattering regimes, which we will briefly discuss next.

3.3.2 Rayleigh Scattering

Rayleigh scattering occurs when the particles are small, on the scale of λ. The mean scattered power from N Rayleigh scatterers in the direction θ (Figure 3.2) is proportional to $\omega^4\alpha^2 E_0^2 \sin^2\theta/32\pi^2\varepsilon_0 rc^3$ (i.e., proportional to ω^4, and so the scattered power is proportional to $1/\lambda^4$).

It is understood that the primary model for describing the field from "small" scattering structures is that of a dipole field (recall Sihvola's dipolarizability (Sihvola, 2007)). For a dipole, the field can be written as follows:

$$E = \frac{1}{4\pi\varepsilon_0}\left[(1 + ik_0 r)\frac{3(\bar{r}\cdot\bar{p})\bar{r} - r^2\bar{p}}{r^2} + k_0^2\frac{r^2\bar{p} - (\bar{r}\cdot\bar{p})\bar{r}}{r^3}\right]e^{-ik_0 r} \qquad (3.26)$$

where $\bar{p}$ is the dipole moment. For a magnetic dipole, we can just replace $\bar{p}$ by $\bar{m}$, the magnetic moment and replace ε_0 by μ_0. Dropping the terms that fall off fast with r and rewriting this in polar coordinates, we can simplify this to give:

$$E_r = \frac{2p\cos\theta}{4\pi\varepsilon_0 r^3}(1 + ik_0 r)e^{-ik_0 r} \qquad (3.27)$$

This is the same as the expression obtained previously when calculating the net phase retardation from a sheet of dipoles.

Other important definitions discussed previously can be extended by substituting σ in the previous scattering equations with C for cross sections. Using this substitution, the scattering cross section is now defined as

$$C_{sc} = \frac{1}{k^2}\int F(\theta,\phi)d\Omega \qquad (3.28)$$

where $d\Omega = \sin\theta\, d\theta\, d\phi$, F is dimensionless, and F/k^2 is the area. Similarly, we can define C_{abs} as the absorption cross section. As previously shown, we can now define the extinction cross section as

$$C_{ext} = C_{sc} + C_{abs} \qquad (3.29)$$

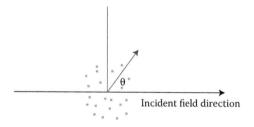

Figure 3.2   Rayleigh scattering definition of variables.

We can now define an "efficiency" factor as

$$Q_{ext} = \frac{C_{ext}}{G} \tag{3.30}$$

where G is the geometrical cross section.

Similarly, we can now define the corresponding efficiency factors Q_{sc}, Q_{abs} and Q_{ext} ($Q_{ext} = Q_{sc} + Q_{abs}$). Extinction from a single particle, Q_{ext}, can be very large, especially at resonance. Defining N as the number of particles per unit volume, we can express the relationship

$$Q_{ext} = NQ_{ext(single\ particle)} \tag{3.31}$$

3.3.3 Mie Scattering

One very specialized and important scenario of the inverse scattering problem that will be looked at in detail later in this book involves a target that is an isotropic, homogeneous, dielectric sphere in 3-D or a disk in 2-D. This type of problem is called a Lorenz–Mie scattering type, which was first published in 1908. The basic setup with variable definitions is shown in Figure 3.3. In general, Maxwell's equations are solved in spherical coordinates utilizing the separation of variables method. The problem is solved for the case when the field is determined at a distance that is much larger than the wavelength, otherwise known as the far-field condition or zone. In the far field, the solution can be expressed in terms of two scattering functions as follows (van de Hulst, 1957):

$$S_1(\Theta) = \sum_{n=1}^{\infty} \frac{2n+1}{n(n+1)} [a_n \pi_n(\cos\Theta) + b_n \tau_n(\cos\Theta)] \tag{3.32}$$

$$S_2(\Theta) = \sum_{n=1}^{\infty} \frac{2n+1}{n(n+1)} [b_n \pi_n(\cos\Theta) + a_n \tau_n(\cos\Theta)] \tag{3.33}$$

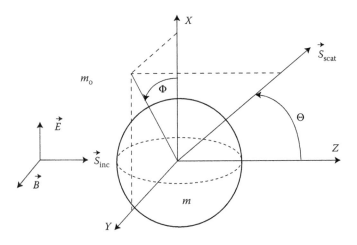

Figure 3.3 Typical setup and coordinate geometry with variable definitions for the Lorenz–Mie scattering problem.

where Θ is the scattering angle as defined in Figure 3.3. Furthermore,

$$\pi_n(\cos\Theta) = \frac{1}{\sin\Theta} P_n^1(\cos\Theta) \tag{3.34}$$

$$\tau_n(\cos\Theta) = \frac{d}{d\Theta} P_n^1(\cos\Theta) \tag{3.35}$$

where P_n^1 represents the associated Legendre polynomials of the first kind. In addition, we have

$$a_n = \frac{\psi_n'(m\alpha)\psi_n(\alpha) - m\psi_n(m\alpha)\psi_n'(\alpha)}{\psi_n'(m\alpha)\xi_n(\alpha) - m\psi_n(m\alpha)\xi_n'(\alpha)} \tag{3.36}$$

$$b_n = \frac{m\psi_n'(m\alpha)\psi_n(\alpha) - \psi_n(m\alpha)\psi_n'(\alpha)}{m\psi_n'(m\alpha)\xi_n(\alpha) - \psi_n(m\alpha)\xi_n'(\alpha)} \tag{3.37}$$

The size parameter, α, is defined as follows:

$$\alpha = \frac{2\pi a m_0}{\lambda_0} \tag{3.38}$$

Finally, in Equations 3.36 and 3.37, the Ricatti–Bessel functions are defined as

$$\psi_n(z) = \left(\frac{\pi z}{2}\right)^{1/2} J_{n+1/2}(z) \tag{3.39}$$

$$\xi_n(z) = \left(\frac{\pi z}{2}\right)^{1/2} H_{n+1/2}(z) = \psi_n(z) + iX_n(z) \tag{3.40}$$

$$X_n(z) = -\left(\frac{\pi z}{2}\right)^{1/2} Y_{n+1/2}(z) \tag{3.41}$$

where Y_n is a typical Bessel function of the second kind.

In addition to these relationships, there are other characteristics of Mie scatterers that are of great interest in this book, in particular the efficiency factor of scattering, which can be approximated by (Walstra, 1964)

$$Q = 2 - \frac{4}{p}\sin(p) + \frac{4}{p^2}(1 - \cos(p)) \tag{3.42}$$

where

$$p = \frac{4\pi r(n - 1)}{\lambda} \tag{3.43}$$

Equations 3.42 and 3.43 are very useful in that they are easily implemented and approximate the scattering cross section predicted by the Lorenz–Mie theory to within 1% (Mohlenhoff et al. 2005; Walstra, 1964).

In a later chapter, the "Q" factor will be utilized in explaining a cyclic phenomenon associated with the performance of reconstructed images obtained from the Born approximation method. It will be shown that resonance, or lack thereof, plays a vital role in the ability to successfully reconstruct an image using the techniques described in this book.

The important and valuable feature of Mie scattering is that one can exactly know the resonances and all one needs for this is the index or (ε_r, μ_r) and $x = 2\pi a/\lambda$ (a is sphere radius, λ is wavelength in the external medium). Mie scattering is rigorous for spheres and other convex shapes. Only the index m and a/λ are more important. One can typically expect strong resonances, especially for larger $|m|$.

If we recall that the polarizability of a scattering atom is expressed as

$$\bar{P} = \alpha \bar{E} \tag{3.44}$$

where $a = \sum a_n \cos w_n t$ (where ω_n is the Mie resonant frequency of the nth mode). In addition, consideration should be given to other contributions to α:

$$\alpha = \alpha_e + \alpha_a + \alpha_d + \alpha_s + \cdots \tag{3.45}$$

where α_e is the electronic component, α_a is the atomic component, α_d is the orientational component, α_s is the shape component, etc.

The next logical question to explore is whether we can relate the effective refractive index of a material to its dipolar or Mie scattering. The scattered field from one particle is

$$u = S(\theta,\phi)u_0 \frac{e^{-ikr}e^{ikz}}{ikr} \tag{3.46}$$

where $S(\theta, \phi)$ is the individual particle scattering pattern and u_0 is the incident wave, $e^{-ikz+i\omega t}$. In the forward direction, the total field is

$$u \simeq u_0\left[1 + S(0)\sum \frac{1}{ikr}e^{-(ik(x^2+y^2)/2r)}\right] \tag{3.47}$$

$$\simeq u_0\left[1 - \frac{2\pi}{k^2} NLS(0)\right] \tag{3.48}$$

which is the approximate expression for total field from integrating over a slab of length L.

If we replace the medium by an equivalent medium having complex index profile, $\tilde{m}$, and write $e^{-ikL(\tilde{m}-1)} \simeq 1 - ikL(\tilde{m} - 1)$, then we can write

$$\tilde{m} = 1 - iS(0)\frac{2\pi N}{k^3} = n - in' \tag{3.49}$$

where $n = 1 + (2\pi N/k^3)\text{Im}(S(0))$, n being the refractive index. It is important to note that the imaginary part determines the overall phase lag or advance for a slab of N particles. Similarly, n' representing the absorption or the decrease in intensity is given by $n' = (2\pi N/k^3)\text{Re}\{S(0)\}$. Using these definitions for refractive index, we can now apply this to the absorption coefficient $\Gamma = 2kn'$ so that $\Gamma(4\pi N/k^2)\text{Re}\{S(0)\} = NC_{\text{ext}}$.

REFERENCES

Barber, P. W. and Yeh, C. 1975. Scattering of electromagnetic waves by arbitrarily shaped dielectric bodies. *Applied Optics*, *14*, 2864–2872.

Keller, J. B. 1961. Backscattering from a finite cone – Comparison of theory and experiment. *IRE Transactions on Antennas and Propagation*, *9*, 411–412.

Logan, N. A. 1965. Survey of some early studies of the scattering of a plane wave by a sphere. *IEEE Proceedings*, *53*, 773–785.

Mohlenhoff, B., Romeo, M., Diem, M., and Wood, B. 2005. Mie-type scattering and non-Beer–Lambert aborption behavior of human cells in infrared microspectroscopy. *Biophysical Journal*, *88*(5), 3635.

Sihvola, A. 2007. Dielectric polarization and particle shape effects. *Journal of Nanomaterials*, article ID 45090, 9 pages.

van de Hulst, H. C. 1957. *Light Scattering by Small Particles*. New York: Wiley.

Walstra, P. 1964. Approximation formulae for the light scattering coefficient of dielectric spheres. *British Journal of Applied Physics*, *15*, 1545.

Four

Inverse Scattering Fundamentals

4.1 CATEGORIZATION OF INVERSE SCATTERING PROBLEMS

In general, inverse scattering problems can be placed in one of the two categories:

1. Weak scattering
2. Nonweak or strong scattering

Weak scattering occurs when the incident wave is only "scattered" once, and this incident wave basically undergoes very little perturbation as it travels through or interacts with the target. This is significant in that in most of these cases the wave inside the target can generally be approximated as the incident wave, which allows the problem to be linearized in order to find a solution. This approximation is what is exploited in the approach to the solution used in the Born and Rytov approximation methods (Avish and Slaney, 1988; Lin and Fiddy, 1990). In these approaches, by linearizing the problem, one is able to establish a Fourier relationship between the measured scattered field data and the target or scattering function. In principle, these methods are only supposed to work well with weak scatterers due to the dependence on this approximation. More precisely, the Rytov approximation requires targets whose permittivity or index varies only very slowly on the scale of the wavelength. This concept of weak scattering will be examined, and in some sense challenged, in this research to understand the extent to which these methods actually work. In the case of strong scatterers, the field inside the target is scattered multiple times and can incur significant perturbation that introduces nonlinearities to the integral equation of scattering (Chew, 1995). This negates the linear approach to this problem, making it quite difficult to solve if and when the above-mentioned methods do not perform well. A solution for the problem involving strong scattering is highly desired, as most targets of interest in real life would fall into this category.

4.2 INVERSE SCATTERING IN TWO DIMENSIONS

A common setup for a 2-D inverse scattering experiment is shown in Figure 4.1. Typically, there is some fixed number of receivers or receiver locations set up in some configuration around the center of the target, usually equally spaced. An incident plane wave then illuminates the target at some known angle ϕ_{inc} with respect to the x-axis and the scattered field is measured at an angle of ϕ_s in the far field. Now, a target or scattering object represented by $V(r)$ is placed in a homogeneous background, which has a permittivity of ε_0, where

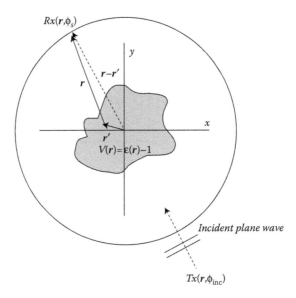

Figure 4.1 A typical 2-D inverse scattering experimental setup. Transmitter Tx transmits the incident quasi-monochromatic plane wave to the scattering object $V(r)$. The receivers Rx are located all around the target which collects the scattered field data after the interaction of the incident wave with the scattering object.

ε_0 is usually the permittivity of free space. This target or scattering object has a permittivity of $\varepsilon(r)$, which is related to the target by the equation

$$V(r) = \varepsilon_r(r) - 1 \qquad (4.1)$$

$V(r)$ is an expression that describes the fluctuations of the permittivity relative to free space in terms of the coordinate system described in Figure 4.1 where $r = (x,y)$. Examining this relationship outside of the target boundary yields the following:

$$V(r) = \varepsilon_r(r) - 1 = \varepsilon_{r_0} - 1 = 1 - 1 = 0 \qquad (4.2)$$

This demonstrates that $V(r)$ is zero at all points outside of the target boundary which has a compact support domain of D, which implies that the Fourier transform of $V(r)$ is an entire function which is completely determined by its exact values on some region in Fourier or k-space from which it could, in principle, be determined everywhere using analytic continuation (Ritter, 2012). The incident plane wave (in the absence of the target) is governed by the scalar homogeneous Helmholtz equation (Avish and Slaney, 1988) as given here:

$$(\nabla^2 + k^2)\Psi_{\text{inc}}(r) = 0 \qquad (4.3)$$

where k is the wave number as defined by $k = 2\pi/\lambda$. The solution for the incident field, Ψ_{inc}, can be written in terms of the standard exponential form of a plane wave

$$\Psi_{\text{inc}}(r) = e^{ik\hat{r}_{\text{inc}} \cdot r} \qquad (4.4)$$

where $\hat{r}_{inc}$ is the unit vector that specifies the direction of the incident field. An expression for the total field $\Psi(r)$ in terms of the coordinate system r can be obtained by looking at the interaction of the incident field, $\Psi_{inc}(r)$, and the target $V(r)$. This relationship can be expressed as an inhomogeneous Helmholtz equation (Avish and Slaney, 1988) as follows:

$$(\nabla^2 + k^2)\Psi(r) = -k^2 V(r)\Psi(r) \tag{4.5}$$

The total field in Equation 4.5 can now be expressed generally as the sum of the incident field and the scattered field as follows:

$$\Psi(r) = \Psi_{inc}(r) + \Psi_s(r) \tag{4.6}$$

where Ψ_s is the scattered field. All fields in Equation 4.6 are expressed in terms of the position as defined by the point r. Now the total field, $\Psi(r)$, can be expressed in terms of an inhomogeneous Fredholm integral equation of the first kind (Morse and Feshbach, 1953) as follows:

$$\Psi(r,\hat{r}_{inc}) = \Psi_{inc}(r) - k^2 \int_D V(r')\Psi(r',\hat{r}_{inc})G_0(r,r')dr' \tag{4.7}$$

where $G_0(r, r')$ is (free space) Green's function which is the solution of the scalar Helmholtz equation (Equation 4.4), and it satisfies the differential equation (Morse and Feshbach, 1953)

$$(\nabla^2 + k^2)G_0(r,r') = \delta(r - r') \tag{4.8}$$

Green's function basically gives the field amplitude at any point r, generated by any given point source located at r'. Since the modeling space in this case is homogenous and rotationally symmetrical, Green's function can be solved in spherical coordinates with r' at the origin as follows:

$$(\nabla^2 + k^2)G_0(r) = -\delta(r) = -\delta(x)\delta(y)\delta(z) \tag{4.9}$$

For homogenous, spherically symmetrical partial differential equations (PDEs), the solution for Green's function in free space can be written as

$$G_0(r) = \frac{Ce^{ikr}}{r} + \frac{Ue^{-ikr}}{r} \tag{4.10}$$

Utilizing the radiation boundary condition, which is simply that the sources cannot be present at infinity, translates to this: that only the outgoing wave solution(s) exist(s) and therefore U must be equal to 0 ($U = 0$). This simplifies Equation 4.10 as follows:

$$G_0(r) = \frac{Ce^{ikr}}{r} \tag{4.11}$$

Equation 4.11 can now be substituted back into Equation 4.9, and the entire equation can then be integrated over a small volume ∇V about the origin as shown:

$$\int_{\nabla V} dV(\nabla^2 + k^2)\frac{Ce^{ikr}}{r} = -\delta(\mathbf{r}) = \int_{\nabla V} dV(-\delta(x)\delta(y)\delta(z)) \tag{4.12}$$

where the right-hand side can be simplified due to the identity of the delta function as follows:

$$\int_{\nabla V} dV(\nabla^2 + k^2)\frac{Ce^{ikr}}{r} = -1 \tag{4.13}$$

Now the left-hand side of Equation 4.13 can be simplified by applying the distributive property as follows:

$$\int_{\nabla V} \nabla^2 \frac{Ce^{ikr}}{r} dV + \int_{\nabla V} k^2 \frac{Ce^{ikr}}{r} dV = -1 \tag{4.14}$$

As V becomes smaller and smaller, the second term in Equation 4.14 gradually goes to zero due to the fact that dV in this term is defined as

$$dV = r^2 \sin\theta \, dr \, d\theta \, d\phi \tag{4.15}$$

and as V gets smaller so does the r^2 term in Equation 4.15 which causes the second term to go to zero as stated earlier. This further simplifies Equation 4.14 as follows:

$$\int_{\nabla V} \nabla^2 \frac{Ce^{ikr}}{r} dV = -1 \tag{4.16}$$

Equation 4.16 can then be rearranged as follows:

$$\int_{\nabla V} \nabla^2 \frac{Ce^{ikr}}{r} dV = \int_{\nabla V} dV \nabla \cdot \left(\nabla \frac{Ce^{ikr}}{r}\right) = -1 \tag{4.17}$$

This now allows the divergence theorem to be applied to Equation 4.17 as follows:

$$\int_{\nabla V} dV \nabla \cdot \left(\nabla \frac{Ce^{ikr}}{r}\right) = \oiint_{S} \nabla \frac{Ce^{ikr}}{r} dS = -1 \tag{4.18}$$

which can be expanded as

$$\oiint_{S} \left(\frac{\partial}{\partial r}\frac{Ce^{ikr}}{r} dS\right) r^2 \sin\theta \, d\theta \, d\phi = -1 \tag{4.19}$$

so that if we now utilize the fact that as r approaches zero, $4\pi C = -1$, which can be used to solve for C as follows:

$$C = \frac{1}{4\pi} \tag{4.20}$$

This value of C can now be substituted back into Equation 4.11, which yields

$$G_0(r) = \frac{e^{ikr}}{4\pi r} \tag{4.21}$$

This can now be generalized by shifting the source back to r' which gives Green's function in the general form for this application as

$$G_0(r, r') = \frac{e^{ik|r-r'|}}{4\pi |r - r'|} \tag{4.22}$$

Equation 4.22 gives Green's function in the form necessary to solve the inhomogeneous Helmholtz equations. In this book, all of the experiments are done in 2-D; therefore, the 2-D form of Green's function can be expressed in terms of a zero-order Hankel function of the first kind (Chew, 1995; Darling, 1984) as follows:

$$G(r, r') = \frac{-i}{4} H_0^{(1)}[k|r - r'|] \tag{4.23}$$

Now, an asymptotic assumption can be applied in that as $r \to \infty$, the zero-order Hankel function of the first kind becomes

$$H_0^{(1)}[k|r - r'|] = \frac{4}{i} \frac{1}{\sqrt{8\pi kr}} e^{i(kr+\pi/4)} e^{-ik\hat{r} - |r-r'|} \tag{4.24}$$

where $\hat{r}' = r/|r|$. Substituting Equation 4.24 back into Equation 4.23, the approximated 2-D Green's function now becomes

$$G(r, r') \cong -\frac{1}{\sqrt{8\pi kr}} e^{i(kr+\pi/4)} e^{-ik\hat{r} - |r-r'|} \tag{4.25}$$

Finally, a general equation for the scattered field can be obtained by substituting Equation 4.25 into Equations 4.6 and 4.7, which gives the following:

$$\Psi_s(\hat{r}, \hat{r}_{inc}) = \frac{1}{\sqrt{8\pi kr}} e^{i(kr+\pi/4)} k^2 \int_D V(r') e^{-ik\hat{r} - |r-r'|} \Psi(r', \hat{r}_{inc}) dr' \tag{4.26}$$

where $\hat{r}_{inc} = (\cos\phi_{inc}, \sin\phi_{inc})$ is the unit vector that describes the direction of the incident plane wave and $\hat{r} = (\cos\phi, \sin\phi)$ is the unit vector that describes the direction of the scattered wave. This equation is of the form of a Fredholm integral equation of the first kind. In order to solve this equation for $V(r)$, we must know or be able to adequately approximate what $\Psi(r)$ is inside the target domain D.

4.3 FIRST BORN APPROXIMATION

One of the more well-known and common approaches in imaging from scattered fields using the diffraction tomography method is the Born approximation. In general, this method assumes that the target is a weakly scattering object, and it is generally used in conjunction with the data inversion method described in the previous section. Many current inverse scattering algorithms utilize this approach even when it is not strictly valid to do so (Avish and Slaney, 1988; Lin and Fiddy, 1990). As already indicated, when using this approach, the problem is linearized and establishes a Fourier relationship between the measured scattered field and some function of the target's scattering properties from which we hope to compute an image. A brief introduction to the first Born approximation is given here.

Recalling from Equation 4.6 that the total measured scattered field at the receivers can be generally expressed as

$$\Psi(r) = \Psi_{\text{inc}}(r) + \Psi_s(r) \tag{4.27}$$

this can be expanded in terms of an inhomogeneous Fredholm integral equation of the first kind (Morse and Feshbach, 1953) as follows:

$$\Psi(r, \hat{r}_{\text{inc}}) = e^{ik\hat{r}_{\text{inc}} \cdot r} + \frac{1}{\sqrt{8\pi kr}} e^{i(kr+\pi/4)} k^2 \int_D V(r') e^{ik\hat{r} \cdot r'} \Psi(r', \hat{r}_{\text{inc}}) dr' \tag{4.28}$$

The first term in the above equation is the contribution from the incident (or illuminating) wave. This term can be premeasured, that is, data obtained when no target is present, and subtracted out, which leaves only the second term in above equation which is the scattered field expressed as follows:

$$\Psi_s(r, \hat{r}_{\text{inc}}) = \frac{1}{\sqrt{8\pi kr}} e^{i(kr+\pi/4)} k^2 \int_D V(r') e^{ik\hat{r} \cdot r'} \Psi(r', \hat{r}_{\text{inc}}) dr' \tag{4.29}$$

$$\Psi_s(r, \hat{r}_{\text{inc}}) = \frac{1}{\sqrt{8\pi kr}} e^{i(kr+\pi/4)} f(k\hat{r}, k\hat{r}_{\text{inc}}) \tag{4.30}$$

where $f(k\hat{r}, k\hat{r}_{\text{inc}})$ is the scattering amplitude, which is defined as

$$f(k\hat{r}, k\hat{r}_{\text{inc}}) = k^2 \int_D V(r') e^{ik\hat{r} \cdot r'} \Psi(r', \hat{r}_{\text{inc}}) dr' \tag{4.31}$$

The basis for the Born approximation is that the total field, Ψ, inside the target can be approximated by the incident field in the above integral in Equation 4.31 as follows:

$$\Psi(\boldsymbol{r}, \hat{\boldsymbol{r}}_{\text{inc}}) = e^{ik\hat{r}_{\text{inc}} - r} \tag{4.32}$$

In theory, for the first Born approximation to be valid (i.e., for an object to be classified as a weak scatterer) a necessary condition is that the product of the target's permittivity, its characteristic dimension, and wave number should be much less than unity (Li and Wang, 2010). This is typically expressed mathematically as

$$\left| kV(\boldsymbol{r})a \right| \ll 1 \tag{4.33}$$

where k is the wave number; the object and permittivity are related by $V(\boldsymbol{r}) = \varepsilon_r(\boldsymbol{r}) - 1$ where, in this example, $\varepsilon_r(\boldsymbol{r})$ is the relative permittivity of the target and 'a' specifies some measure of the physical size of the target. It should be noted that the absolute value of this product should be less than 1 to account for new meta-materials that may have a permittivity that is negative. It is widely expected and believed by many that as the dimensions of the object increase, or the magnitude of absolute value of the permittivity fluctuations increases, the first Born approximation becomes increasingly poor. While this does seem to be the case in general, this assumption will be examined in more detail, and the limits of this will be better defined later in this book.

In order to examine the Born approximation in more detail, Equation 4.32 can be substituted into Equation 4.31 which yields the linearized version of the inversion problem as follows:

$$f^{\text{BA}}(k\hat{\boldsymbol{r}}, k\hat{\boldsymbol{r}}_{\text{inc}}) = k^2 \int_D V(\boldsymbol{r}')e^{-ik(\hat{r} \cdot \hat{r}_{\text{inc}}) \cdot r'}\mathrm{d}r' \tag{4.34}$$

which illustrates where the Ewald circles (introduced later) originate and where the scattered field within the first Born approximation is expressed as

$$\Psi_s^{\text{BA}}(\boldsymbol{r}, \hat{\boldsymbol{r}}_{\text{inc}}) = \frac{1}{\sqrt{8\pi kr}} e^{i(kr+\pi/4)}k^2 \int_D V(\boldsymbol{r}')e^{-ik(\hat{r} \cdot \hat{r}_{\text{inc}}) \cdot r'}\mathrm{d}r' \tag{4.35}$$

Equation 4.35 gives the Fourier relationship between the target or scattering object, $V(\boldsymbol{r})$, and the measured scattering amplitude at the receivers $f^{\text{BA}}(k\hat{\boldsymbol{r}}, k\hat{\boldsymbol{r}}_{\text{inc}})$. If the integral in Equation 4.35 is isolated as follows

$$\int_D V(\boldsymbol{r}')e^{-ik(\hat{r} \cdot \hat{r}_{\text{inc}}) \cdot r'}\mathrm{d}r' = \frac{\left(\sqrt{8\pi kr}\right)e^{-i(kr+\pi/4)}}{k^2} \Psi_s^{\text{BA}}(\boldsymbol{r}, \hat{\boldsymbol{r}}_{\text{inc}}) \tag{4.36}$$

the right-hand side of Equation 4.36 is what is implemented in MATLAB® code later in this text. This gives a representation not of $V(\boldsymbol{r})$, but it gives an estimate for the product $V(\boldsymbol{r})\Psi/\Psi_{\text{inc}}$. The challenge is now to determine how to identify Ψ and effectively remove or minimize it to recover a valid representation of the target, $V(\boldsymbol{r})$.

One challenge with the Born approximation is the inconsistency mentioned by Ramm (1990). Ramm demonstrated that the target or scattering object $V(\boldsymbol{r})$

is real if and only if its Fourier transform is Hermitian. This means that in order for $V(\mathbf{r})$ to be real, it must be equal to the transpose of its complex conjugate. This is expressed as

$$-\frac{i}{2}\left\{ f^{\text{BA}}(k\hat{\mathbf{r}},k\hat{\mathbf{r}}_{\text{inc}}) - \left[f^{\text{BA}}(k\hat{\mathbf{r}},k\hat{\mathbf{r}}_{\text{inc}}) \right]* \right\} = \text{Im}\left[f^{\text{BA}}(k\hat{\mathbf{r}}, k\hat{\mathbf{r}}_{\text{inc}}) \right] \qquad (4.37)$$

which can also be written as

$$\frac{k}{4\pi} \int\limits_{\Omega \in H^2} d\Omega \left| f^{\text{BA}}(k\hat{\mathbf{r}},k\hat{\mathbf{r}}_{\text{inc}}) \right|^2 = 0 \qquad (4.38)$$

where Ω is defined as the solid angle over which H^2 is integrated in either 2-D or 3-D. Equation 4.38 clearly suggests that the target must be zero everywhere for the Born approximation to be valid. This simply is not possible for a real scattering object. Therefore, this indicates that there can be no exact solution for real objects for the Born approximation. Physically, $V(\mathbf{r})$ is always complex, as dictated by dispersion relations that follow directly for causal systems. This notwithstanding, for sufficiently small scattering objects with negligible noise levels, a consistent stable estimate of $V(\mathbf{r})\Psi/\Psi_{\text{inc}}$ can be obtained by using a regularized inversion of the Fourier data (Ritter, 2012), such as the discrete (inverse) Fourier transform (DFT).

Taking another look at Equation 4.34, it is evident that the inverse Fourier transform of the complex far-field scattering amplitude can provide a reasonable representation of the scattering object. As already discussed in the previous section, this can be accomplished with one of the two methods, namely, the Fourier transform-based interpolation and the filtered back propagation method (Avish and Slaney, 1988). When the target or scattering object is considered to be a nonweak or a strong scatterer, theoretically the Born approximation is said to be invalid. When this is the case, the Fourier inversion of many Ewald circles discussed in the next chapter results in an approximation of $V(\mathbf{r})$, which is expressed as

$$V_{\text{BA}}(\mathbf{r},\hat{\mathbf{r}}_{\text{inc}}) \approx V(\mathbf{r})\left\langle \frac{\Psi(\mathbf{r},\hat{\mathbf{r}}_{\text{inc}})}{\Psi_{\text{inc}}(\mathbf{r},\hat{\mathbf{r}}_{\text{inc}})} \right\rangle \qquad (4.39)$$

In this equation, the angle bracket symbols, $\langle\,\rangle$, indicate an averaged dependence on the direction of the incident plane wave. The form of this equation suggests that the Ψ-averaged term is independent of the $V(\mathbf{r})$ term, which, if true, implies that it should be separable. It is unclear at this time if this approximation, that is, that these terms are independent of each other, holds for all classes of targets, but we can argue that for increasingly complex targets, the averaged internal multiplied scattered fields will become increasingly noise-like, while $V(\mathbf{r})$ remains unchanged. Equation 4.39 is an approximation because, in principle, the Fourier transformation can only be applied for each $\hat{\mathbf{r}}_{\text{inc}} = \text{constant}$, and also, in practice, its accuracy will be affected by the limited data (i.e., k-space) covered, which is always the case in this text.

4.4 RYTOV APPROXIMATION

While the Born approximation is one of the more commonly used methods for linearizing an inverse scattering problem, there are other methods available. One of the more common alternative methods is the Rytov approximation. In this, the total field is represented in terms of a complex phase (Avish and Slaney, 1988; Ishimaru, 1978) shown in Equation 4.40:

$$\Psi(r, \hat{r}_{inc}) = \Psi_{inc}(r)e^{i\Phi_s(r)} \tag{4.40}$$

where $\Phi(r)$ is the complex phase function and is defined as

$$\Phi(r) = \Phi_{inc}(r) + \Phi_s(r) \tag{4.41}$$

In Equation 4.41, the $\Phi_s(r)$ term is the phase function of the scattered field. If this new representation of the total field in Equation 4.40 is now substituted back into the general inhomogeneous Helmholtz equation (Equation 4.5) and the defined identity functions are used, then this yields the following:

$$\nabla^2[\Psi_{inc}(r)\Phi_s(r)] = [\nabla^2\Psi_{inc}(r)]\Phi_s(r) + 2\nabla\Psi_{inc}(r) \cdot \nabla\Phi_s(r) + [\nabla^2\Phi_s(r)]\Psi_{inc}(r) \tag{4.42}$$

Simplifying, the inhomogeneous Helmholtz equation then becomes (Lin and Fiddy, 1990)

$$(\nabla^2 + k^2)[\Psi_{inc}(r)\Phi_s(r)] = i\{k^2V(r) - [\nabla\Phi_s(r)]^2\}\Psi_{inc}(r) \tag{4.43}$$

As in Chapter 2, we utilize the free-space Green's function in Equation 4.43; the complex phase function can then be written as

$$\Phi_s(r) = \frac{ik^2}{\Psi_{inc}(r)}\int_D V(r')\Psi_{inc}(r')G_0(r,r')dr'$$

$$- \frac{i}{\Psi_{inc}(r)}\int_D [\nabla\Phi_s(r)]^2\Psi_{inc}(r')G_0(r,r')dr' \tag{4.44}$$

Interestingly, in Equation 4.44, the second term can be approximated to zero if the following criteria are met (Lin and Fiddy, 1990)

$$|k^2V(r)| \gg |([\nabla\Phi_s(r)]^2)| \tag{4.45}$$

$$|\varepsilon(r) - 1| \gg \left|\left(\left[\frac{\nabla\Phi_s(r)}{2\pi}\right]^2\right)\right| \tag{4.46}$$

The implication or impact of these inequalities is that in order for these criteria to be met, the incident wavelength must be very small in comparison to the mean size of the target or scattering object, or that the spatial rate of phase change induced by the target or scattering object must be very small in terms of the unit wavelength. Equation 4.46 implies that this approximation begins to break down as $V(r)$ approaches zero. A global condition for the validity of the Rytov approximation is

$$\left| k^2 \int_D V(r')\Psi_{inc}(r')G_0(r,r')dr' \right| \gg \left| \int_D [\nabla\Phi_s(r)]^2\Psi_{inc}(r')G_0(r,r')dr' \right| \quad (4.47)$$

So, when this criterion is satisfied and the Rytov approximation is valid, the complex scattered phase can be expressed as

$$\Phi_s(r) = \frac{ik^2}{\Psi_{inc}(r)} \int_D V(r')\Psi_{inc}(r')G_0(r,r')dr' \quad (4.48)$$

This can now be substituted back into Equation 4.40 to compute the total field as follows:

$$\Psi(r,\hat{r}_{inc}) = \Psi_{inc}(r)e^{i\left[(ik^2/\Psi_{inc}(r))\int_D V(r')\Psi_{inc}(r')G_0(r,r')dr'\right]} \quad (4.49)$$

If the argument of the exponent is now isolated by dividing by the incident field and the logarithm applied, the resulting form of this equation is

$$\Psi_{inc}(r,\hat{r}_{inc})\ln\left[\frac{\Psi(r,\hat{r}_{inc})}{\Psi_{inc}(r,\hat{r}_{inc})}\right] = -k^2 \int_D V(r')\Psi_{inc}(r')G_0(r,r')dr' \quad (4.50)$$

This equation is comparable to Equation 4.35 from the Born approximation analysis in that it basically defines an inverse Fourier relationship or procedure to recover $V(r)$. Equation 4.50 can be very difficult to evaluate due to the nature and challenges of dealing with the multivalued issues of the natural logarithm (Fiddy et al., 2004). We will encounter the same difficulty in Chapter 8.

Also, comparable with the Born approximation, when the conditions for the Rytov approximation are not valid, $V_{RA}(r)$ is recovered in lieu of $V(r)$ where

$$V_{RA}(r,\hat{r}_{inc}) = V(r) - \frac{1}{k^2}[\nabla\Phi_s(r)]^2 \quad (4.51)$$

REFERENCES

Avish, C. K. and Slaney, M. 1988. *Principles of Computerized Tomographic Imaging.* New York: IEEE Press.

Chew, W. C. 1995. *Waves and Fields in Inhomogeneous Media.* Piscataway: IEEE Press.

Darling, A. M. 1984. *Digital Object Reconstruction from Limited Data Incorporating Prior Information.* Thesis, University of London.

Fiddy, M. A., Testorf, M., and Shahid, U. 2004. Minimum-phase-based inverse scattereing method applied to IPS008. *SPIE, 5562*, 188–195.

Ishimaru, A. 1978. *Wave Propagation and Scattering in Random Media* (Vol. 1). New York: Academic Press.

Li, J. and Wang, T. 2010. On the validity of the Born approximation. *Progress In Electromagnetics Research, 107*, 219–237.

Lin, F. C. and Fiddy, M. A. 1990. Image estimation from scattered field data. *International Journal of Image System Technology, 2*, 76–95.

Morse, P. M. and Feshbach, H. 1953. *Methods of Theoretical Physics.* New York: McGraw-Hill.

Ramm, A. G. 1990. Is the Born approximation good for solving the inverse scattering problem when the potential is small? *Journal of Mathematical Analysis and Applications, 147*(2), 480–485.

Ritter, R. S. 2012. *Signal Processing Based Method for Modeling and Solving Inverse Scattering Problems.* University of North Carolina at Charlotte, Electrical Engineering. Charlotte: UMI/ProQuest LLC.

II

INVERSION METHODS

Five

Data Processing

5.1 DATA INVERSION IN k-SPACE: A FOURIER PERSPECTIVE

Basically, there are two methods or algorithms that are commonly used in performing Fourier-based inversion of scattering field data. They are

1. Fourier-based interpolation
2. Filtered back propagation

Other methods, which are not addressed in this book, are more complex and computationally intensive methods such as the modified gradient techniques that are iterative in nature and do not necessarily converge for strong scatterers. Detailed discussions of these methods can be found in Avish and Slaney (1988). The basic methodology used in the Fourier interpolation method is that the scattered field data are placed on semicircular arcs in 2-D k-space or the Fourier domain where the loci of points are defined by the Ewald circles or, in 3-D, Ewald spheres (Wolf, 1969). The Ewald circles arise naturally from Equation 4.26 when adopting the first Born approximation. They are tangent to the origin of 2-D k-space with a radius of k, where k represents the magnitude of the scattered field's wavenumber. The transmitted data, that is, forward scattered data lying on the part of the circle closest to the origin and the reflected data, that is, backscattered data lying on the part of the circle farthest away from the origin are depicted in Figure 5.1. The position of the circle is relative to the direction of the source of the incident wave that is illuminating the target. As the incident wave is moved or rotated around the target and the subsequent data gathered from the stationary receivers and mapped onto the respective Ewald circle, additional Ewald circles of data are formed in k-space.

Ideally, as the source is rotated all the way around the target, k-space is filled to some degree with data out to a "limiting circle" of radius $2|k|$ as shown in Figure 5.2. This method can be used to develop an estimate of the Fourier transform of the target for the given incident frequency, but theoretically this is only a representative of an image in the weakly scattering limit known as the first Born approximation. For more general scattering targets, these Fourier data must be interpreted in terms of the integral equation given in Equation 4.26 indicating the dependence on both the scattering distribution and the total field that resides inside the scattering volume. Other incident frequencies can be used, which will in turn vary the radius of the corresponding Ewald circle to help fill k-space. The radius of the Ewald circle is directly proportional to the frequency of the incident illuminating source as illustrated in Figure 5.3.

In the filtered back propagation method mentioned above, the scattered field data are "propagated" backwards into the object domain using an appropriate

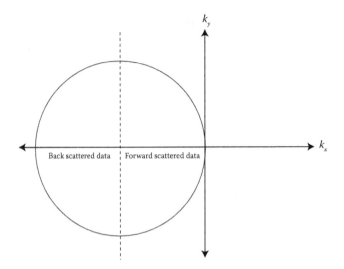

Figure 5.1 Spatial frequency space or *k*-space map showing that the forward scattered field data lies close to the origin and backscattered data lies furthest from the origin assuming that the target is illuminated from the −*x* direction.

Green's function, which for this (and most) cases is assumed to be Green's function for free space (Chew, 1995; Lin and Fiddy, 1990). It is this assumption that also leads to the Fourier transform relationship being exploited as seen elsewhere in this text on the discussion on the Born approximation. However, as mentioned above, the back propagation step generates a field distribution in

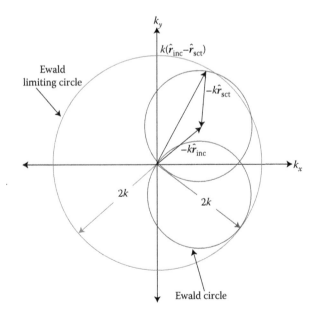

Figure 5.2 Fourier space (*k*-space) of the object as a result of interaction of different incident plane waves with scattering object. The direction of the incident field $\hat{r}_{inc}$ and the direction $\hat{r}_{sct}$ of a particular plane wave component of the scattered field define a point at the Ewald circle. Changing incident field directions $\hat{r}_{inc}$ fills the interior of Ewald limiting circle.

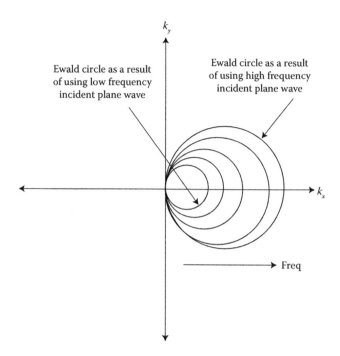

Figure 5.3   The radius of these Ewald circles is specified by the magnitude of the k-vector and so changing the incident frequency and hence, k will change the k-space data mapping as shown.

the target domain that, in theory, is only proportional to the target itself in the weakly scattering limit, that is, in the first Born approximation.

5.2 TARGET MODELING AND DATA GENERATION

The task of 2-D target modeling and image data generation from scattered fields is not a trivial one. The problem lies in that there is no general solution for analytically determining scattered fields for an arbitrary target. There are some analytical solutions only for a few very simple targets, but in general, no solution exists. This means that for the varying case-by-case situations, the analytical solution would have to be derived each time for a new target, if in fact the analytical solution does exist at all in a closed empirical form, which is unlikely. A common numerical solution to this type of problem or modeling is to use the technique of finite element analysis (Jin, 2002; Silvester and Ferrari, 1996). In this method, the differential equations involved in calculating these scattered fields are solved numerically in an iterative process. The basic model setup for this procedure is similar to the general model shown in Figure 1.1, with the exception that there is an artificial boundary that defines the extent that the iterative calculations are performed for, since this is a finite method as shown in Figure 5.4. At this boundary the properties of the boundary are defined such that there are no reflections and it gives the "appearance" that the model space goes on forever. The general solution for an E_z polarized field in the model space satisfies the scalar Helmholtz equation as follows:

$$\frac{\partial}{\partial x}\left(\frac{1}{\mu_r}\frac{\partial E_z}{\partial x}\right) + \frac{\partial}{\partial y}\left(\frac{1}{\mu_r}\frac{\partial E_z}{\partial y}\right) + k_0^2\varepsilon_r E_z = jk_0 Z_0 J_z \qquad (5.1)$$

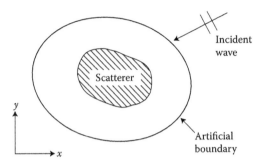

Figure 5.4   The general 2-D finite element scattering model.

This is the basic general equation that is used for the finite element method in the model space that is solved iteratively along the finite element mesh. The formation of this finite element mesh and the derivation and application of the elemental interpolation is beyond the scope of this book, but can be seen in detail in Jin (2002). The bounded area is enclosed using perfectly matched layers (PML) that utilize the general relationships along the mesh on the boundary of

$$\frac{\partial \varphi^{sc}}{\partial \rho} + \left(jk_0 + \frac{1}{2\rho} \right) \varphi^{sc} = 0 \tag{5.2}$$

$$\frac{\partial \varphi^{sc}}{\partial \rho} + \left[jk_0 + \frac{1}{\rho} - \frac{1}{8\rho^2 \left((1/\rho) + jk_0 \right)} \right] \phi^{sc} - \frac{1}{2\rho^2 \left((1/\rho) + jk_0 \right)} \frac{\partial^2 \phi^{sc}}{\partial \varphi^2} \tag{5.3}$$

While this method can be extremely complicated, it can be easily implemented using commercially available finite element software such as COMSOL® to simplify this effort. This method and software packages are used to calculate the total field at each receiver location around the target. This greatly reduces the complexity of the approach to these types of problems, but can be computationally costly. These types of software packages allow the user to create the target graphically, modify and/or sweep virtually any and all parameters of interest; the program then applies the finite element process to the model and returns both a graphical and numerical solution for the total field in the defined space. The only challenge then is to take the output and process the data into a format that can be used by imaging algorithms developed in software packages such as MATLAB®, which can typically be done in a commercially available spreadsheet such as Microsoft® Excel.

The basic COMSOL models developed and used in this text are similar to that shown in Section 4.2. The models in this text are set up as shown in Figure 4.1 with a fixed number of receivers equally spaced at a fixed and common distance from the target origin and a fixed number of source locations equally spaced around the target. This can produce some very high quality data files that can be used to test new and existing imaging algorithms as will be shown later in this text. For the purpose of demonstrating the validity of the data generated using this method, the imaging technique utilizing the Ewald circles technique (Wolf, 1969) along with the Born approximation is used on the data obtained from this modeling method to produce a Fourier image of various targets so that a reconstructed image can be displayed and compared to measured real-life data for identical targets. This method is discussed at length in Ritter (2012) and Shahid et al. (2008).

5.3 TARGET MODELING ENVIRONMENT

The typical model used throughout this text consists of a target area centered about the origin. Receiver points, which are used to "measure" or obtain the calculated complex scattered field values at these points, are located on an imaginary circle centered about the origin with an arbitrarily fixed radius of 760 mm. There are 360 data points equally spaced along this circle, which basically gives the ability to measure the total field on 1° increments around the target in the 2-D plane. The illuminating source is "cycled" or rotated around the target in 36 equally spaced locations, which basically gives the ability to view the scattered data from a source rotated in 10° increments around the target. This translates to having the capability of creating 36 separate Ewald circles as previously discussed in Wolf (1969). As already mentioned, this basic model environment has been successfully implemented, and data has been successfully gathered from the simulations and formatted for use in the already developed MATLAB algorithms used in Shahid (2009) and in this book. The effectiveness of this modeling method will be demonstrated in the next section.

The implementation of this environment in the COMSOL software for a basic cylinder target is shown in Figure 5.5. In this figure, the basic model, the mesh, the Z component of the E-field which is orthogonal to the plane of propagation and the normalized E-field are shown to illustrate the capabilities of the software and model. It is apparent from these images that this modeling technique is functioning properly and should provide valid data. The data obtained from this modeling process is then exported to a Microsoft Excel spreadsheet and processed to be formatted to be used as a data file in MATLAB. This processing of the data basically consists of deleting the "i" terms inserted by COMSOL to designate the imaginary component of the data pairs. MATLAB does not recognize this format and therefore the data are formatted such that every other number is the real portion of the data pairs, while the accompanying every other number is the imaginary component of the measured data pair. This format will be accounted for the MATLAB code.

5.4 IMAGING ALGORITHM IMPLEMENTATIONS: EXAMPLE RECONSTRUCTIONS

In order to demonstrate the validity of this modeling process, the data obtained from the COMSOL model described in the previous section was extracted, similarly processed, and compared to the measured data obtained from the Institut Fresnel (Belkebir and Saillard, 2001, 2005; Geffrin et al. 2005) website for a range of targets. This data was similarly processed using basically the same algorithm described in Ritter (2012) that maps the data for a given source frequency onto Ewald circles and then concurrently applies the Born approximation algorithm to the data to produce a first Born approximation reconstructed image in MATLAB. To demonstrate that the data from the COMSOL/

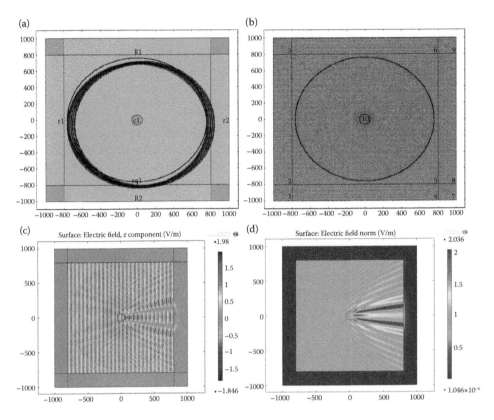

Figure 5.5 (a) Basic model of a circular target (i.e., a cylinder in 3D) in the x–y plane with a radius of 60 mm. The larger circle is the location of the 360 receiver points (with I-D tags shown in the lower right of each receiver point). The square and rectangular sections around the border of the space are matched boundary layers. (b) Basic target model from (a) with finite element mesh applied. (c) Graphical representation of E_z with incident source frequency of 5 GHz for a target in (a) with a relative permittivity of 1.5. (d) Graphical representation of normalized E_z for conditions in (c).

MATLAB modeling process is valid compared to the measured data, the two images for each target, one from the reconstructed images obtained from the measured data in Shahid (2009) and one from the simulated data processed using the same algorithm, respectively, are shown to be comparable in appearance. The target definitions and outputs from both data sets are presented in Table 5.1 for comparison. There are some obvious differences in appearance, which is to be expected to some degree.

One major difference in the data sets for at least the first two targets is that the simulated data was constructed using nine incident sources equally spaced around the target and the measured data was constructed using eight incident sources equally spaced around the target. The last two targets both utilize 18 source locations, again equally spaced around the target. Otherwise, the setups are virtually identical for each target set. While the evaluation of images is in general a subjective task, it should be fairly obvious to even the beginner that the two images are more similar than they are different, suggesting that the structured method described is valid for producing simulated scattered field data from known structures. This is important to understand and accept as we use this method to extensively test imaging methods and ideas in the remainder of this book.

Table 5.1 Institut Fresnel Target Definitions

	Institut Fresnel Data Target Setup	Born Image from Simulated Data	Born Image from Institut Fresnel Measured Data
FoamDielInt			
FoamDielExt			
FoamMetExt			
FoamTwinDiel			

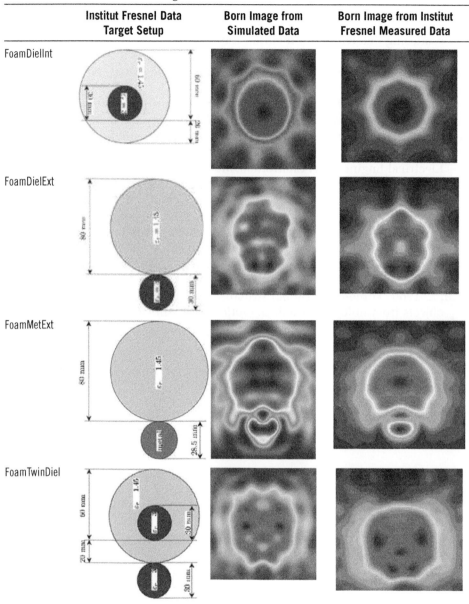

Source: From Belkebir and Saillard (2001, 2005) and Shahid (2009).

REFERENCES

Avish, C. K. and Slaney, M. 1988. *Principles of Computerized Tomographic Imaging.* New York: IEEE Press.

Belkebir, K. and Saillard, M. 2001. Special section on testing inversion algorithms against experimental data, *Inverse Problems*, 17, 1565–1571.

Belkebir, K. and Saillard, M. 2005. Special section on testing inversion algorithms against experimental data: Inhomogeneous targets, *Inverse Problems*, 21, S1–S3.

Chew, W. C. 1995. *Waves and Fields in Inhomogeneous Media.* Piscataway: IEEE Press.

Geffrin, J.-M., Sabouroux, P., and Eyraud, C. 2005. Free space experimental scattering database continuation: Experimental set-up and measurement precision, *Inverse Problems*, 21, S117–S130.

Jin, J. 2002. *The Finite Element Method in Electromagnetics.* New York: IEEE Press.

Lin, F. C. and Fiddy, M. A. 1990. Image estimation from scattered field data. *International Journal of Image System Technology, 2,* 76–95.

Ritter, R. S. 2012. *Signal Processing Based Method for Modeling and Solving Inverse Scattering Problems.* University of North Carolina at Charlotte, Electrical Engineering. Charlotte: UMI/ProQuest LLC.

Shahid, U. 2009. *Signal Processing Based Method for Solving Inverse Scattering Problems.* PhD Dissertation, Optics, University of North Carolina at Charlotte, Charlotte: UMI/ProQuest LLC.

Shahid, U., Fiddy, M. A., and Testorf, M. E. 2008. Inversion of strongly scattered data: Shape and permittivity recovery. *SPIE, 7076,* 606.

Silvester, P. and Ferrari, R. 1996. *Finite Elements for Electrical Engineers.* New York: Cambridge University Press.

Wolf, E. 1969. Three-dimensional structure determination of semi-transparent objects from holographica data. *Optics Communication, 1,* 153–156.

Six

Born Approximation Observations

6.1 DEGREES OF FREEDOM

In the previous discussions, the scattering problem is viewed as an inverse Fourier problem, which is a valid approach, but possibly not a complete one. There is another way of viewing this problem, which may give more insight into what is actually going on and what could possibly give a better criteria for performance. If one were to view one isolated Ewald circle map of the data from most penetrable targets, one would notice that most of the information (that is, nonzero information) is located near the origin or in the forward scattering section of the circle as shown in Figure 6.1. This being the case, as most of the information is in the forward scattering mode one could think of this as the information about the scatterer is "transmitted" through the target. This could mean that another valid approach to this problem would be to treat it as a transmission problem (Miller, 2007) where one hase a source (the incident wave), a transmission medium (the target), and a receiver (the receivers located in the forward half of the Ewald circle). It is known from communication theory (Jones, 1988; Kasap, 2001) or analysis that in order for certain types of signals to be successfully transmitted over a given medium that a minimum number of modes or bandwidth must be present to represent the necessary amount of information at the other end of the medium. This is sometimes referred to as the minimum degrees of freedom necessary for minimum image reconstruction. In this case, the available degrees of freedom are merely a function of the physical characteristics of the medium alone for a given wavelength (Jones, 1988; Kasap, 2001; Miller, 2007). Using this approach and applying it to this application as shown and described in Miller (2007), Kasap (2001), Jones (1988), and Ritter (2012), the general relationship that predicts the minimum degrees of freedom in 3-D is

$$N_{3\text{-D}} = \frac{B_V \cdot n_{\max}}{\lambda^3} \tag{6.1}$$

where $N_{3\text{-D}}$ is the minimum degrees of freedom required in 3-D, B_v is the target volume, $n_{\max}$ is the maximum index of refraction, and λ is the wavelength which can be modified easily for 2-D as follows:

$$N_{2\text{-D}} = \frac{A_V \cdot n_{\max}}{\lambda^2} \tag{6.2}$$

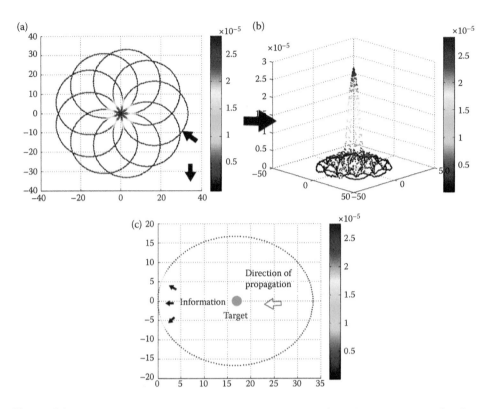

Figure 6.1   (a) Fourier representation from nine Ewald circles obtained from nine separate source directions around a target. (b) Isometric view of Fourier domain image in (a).* (c) One Ewald circle indicated by an arrow in (a) showing direction of incident wave, and resultant location of nonzero receiver data showing comparison to transmission problem as the target or "transmission medium" is located in the center of the Ewald circle.

where the target volume is replaced with the target area A_v. While the criteria mentioned before in the Fourier approach do have merit, it is really more of a measure of the "weakness" of the scatterer and also of the validity of the assumption that the incident field is linear inside the target. The recent challenge is that this new criterion defined above could be a better gauge of the performance or ability of an algorithm to reconstruct or "transmit" the original target. In this application, it is suggested that the degrees of freedom translate to the minimum number of independent target illumination directions or sources, or the minimum number of receivers or some combination thereof necessary to fully "transmit" and "receive" all the information related to the target. A series of experiments will be presented later to illustrate this in the following sections. In particular, each aspect of this criterion will be examined separately. More specifically, the degrees of freedom criteria will be applied to the number of sources first, then, with some modifications, this criterion will be applied to the number of receivers. Later in this book, the idea of an overall degrees of freedom requirement will be examined in more detail. An attempt will be made to demonstrate that the degrees of freedom requirements for the number of sources and receivers spaced uniformly are important. It is possible that these criteria can be combined to form a somewhat

* Note that if the target is a single point, then Figure 6.1b would be the point spread function.

unified degrees of freedom criterion. Experiments are carried out to test this, and speculation is made as to the exact form of this alternate requirement.

6.2 REQUIREMENTS FOR DEGREES OF FREEDOM FOR SOURCES

For the purpose of demonstrating the role of degrees of freedom and its usefulness in predicting performance when imaging from scattered fields, a series or "families" of reconstructed images will now be presented. These reconstructions are presented in a systematic way for a group of selected targets, all illuminated by an increasing number of incident waves, all with a frequency of 5 GHz. The targets of choice for this demonstration are defined and described in Table 6.1.

For each of these target parameters, the degrees of freedom and the weakly scattering metric (kVa), discussed in Section 6.1, are calculated and shown for reference. Also, to aid in the analysis of these images, the borders of each image are shaded in either "red" or "green," with "red" indicating that for the given parameters and number of sources, the minimum degrees of freedom are not satisfied, and the "green" indicating that for the given parameters and number of sources, the minimum degrees of freedom are satisfied. With this color code the reader should be able to distinguish, as the number of sources is increased (moving "down" the table), when the minimum criteria are met.

It should be noted that there are obviously other factors that are affecting the reconstruction in these images, namely obvious resonances that occur for different parameters that definitely affect some of the images as compared to others. These phenomena will be examined in more detail in the following sections. It should also be kept in mind that the measure of and meeting of the number of degrees of freedom do not necessarily indicate that the image will look exactly like the target because the challenges and issues involved with the inverse scattering problem remains; it simply means that for the given parameters, adequate sources have been used to "communicate" or transfer as much information about the target as is possible under the given experimental arrangement. So, with this in mind, one would expect that after the minimum degrees of freedom have been met, and for all images with higher degrees of freedom, the image of the "target" should not change significantly from one image to the next. This is not to say that the entire image will not change as there could be significant differences in the "noise" artifacts in the free space

Table 6.1   Definition of Various Target Types and Scenarios Used to Test the Degrees of Freedom Requirement along with Corresponding Table Number of Results

Table Number	Shapes	Dimensions	Permittivity Range
5.2	1 Circle	Radius = 1λ	2–10
5.3	1 Circle	Radius = 1λ	11–19
5.4	1 Circle	Radius = 2λ	2–10
5.5	2 Circles	Radius = 1λ	1.1–1.9
5.6	2 Circles	Radius = 1λ	2–10
5.7	1 Square	Sides = 1λ	1.1–1.9
5.8	2 Squares	Sides = 1λ	1.1–1.9
5.9	2 Triangles	Base/height = $2\lambda/3\lambda$	1.1–1.9

areas, but that the image of the target itself, no matter how close or far it is in appearance from the original target, should basically remain the same as the number of sources is increased.

Using the criteria described above, it seems evident that images obtained when measurements are below the degree of freedom threshold do not seem to be well formed and do changes from one image to the next, while the images obtained above the degrees of freedom threshold do seem to have a consistent reconstructed image that changes very little as the number of sources increases. This would strongly suggest that the concept of the degrees of freedom presented above for this scenario is valid. This issue will be examined further in relation to the number of receivers in later sections (Tables 6.2 through 6.6).

6.3 REQUIREMENTS FOR DEGREES OF FREEDOM FOR RECEIVERS

The requirement for the minimum number of degrees of freedom demonstrated in the previous section for the number of sources is also similarly applicable for the minimum number of receivers as well. This being the case, it seems appropriate to demonstrate this concept as well on the number of receivers much like those done for the sources. This is not as simple as it might seem. It has been observed that as one continues to reduce the number of receivers and keep the receivers equally spaced and at the same distance from the target, the natural geometrical consequence is that the spacing between the receivers is ever increasing. Eventually, as is already known in 1-D signal processing (Lustig et al., 2007), this will eventually lead to aliasing issues in the resulting reconstructed image. This is exactly what happens in this case. This is a very significant issue, especially dealing with receiver numbers in the 5–20 range for the specific examples investigated here. This issue prevents us from following the direct approach used in the previous section for the sources. Before this aspect of the requirements of minimum degrees of freedom can be sufficiently examined, the issue of aliasing has to be addressed. One approach utilized in 1-D signal processing to address aliasing in undersampled signals is to randomly space the samples in lieu of using equally spaced samples (Lustig et al., 2007). This is a common technique that when employed eliminates or greatly reduces the effects of aliasing, and as long as the original signal strength or signal-to-noise ratio is high enough, the original signal will be evident in the presence of the spatial noise introduced by reducing and randomizing the receiver locations. This process is illustrated in Figure 6.2 taken from Lustig et al. (2007). This same method is the technique that is implemented in 2-D in the current algorithm code in an attempt to disperse the aliasing as done in 1-D. This will of course introduce some element of noise to the reconstructed images, but the signal-to-noise ratio should be sufficient to produce a useful image for examining the effects of the degrees of freedom on the number of receivers.

Since the total number of receivers used in this model is 360, any multiple of 360 can be used for the number of receivers and still maintains the equal spacing of the receivers. This being the case, the MATLAB® code was modified such that for any undersampled quantity of receivers used, that is, less than 360, the user has the ability to use the random generator in MATLAB to

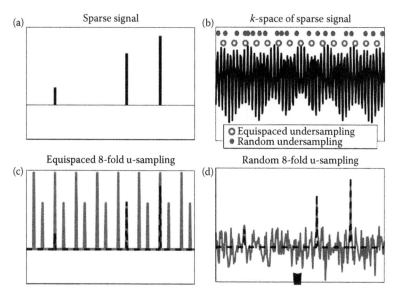

Figure 6.2   (a) Sparse signal to be sampled. (b) The *k*-space representation of original signal in (a) with equispaced undersampling and random undersampling sample locations shown. (c) Resulting reconstruction of original sampled signal using equispaced samples. (d) Resulting reconstruction of original sampled signal using random samples.

have the receiver at any location between the chosen locations to be selected in a Gaussian distribution centered on the original receiver location. This in a sense creates, for lack of a better term, what we define here as a noise bandwidth (NBW) around the chosen receiver location. For example, if 36 receivers are chosen, then the maximum noise bandwidth would be $360/36 = 10$. This means that the receiver location could be any location within $\pm 5°$ around the original receiver location. Naturally, as fewer receivers are used, the NBW can increase if desired.

As the consequences of doing this will be a function of frequency, this approach was initially tested for three different frequencies to observe the effects. The three incident frequencies used are 8, 5 and 2 GHz. The simplest case was modeled first, that being a circle or cylinder with a radius of λ respective of the incident frequency. In conjunction with testing this possible remedy for the aliasing, the degree of freedom criteria was examined as well. For each case in this first series of tests, the minimum number of degrees of freedom is calculated to be $\pi(n)$ and for these tests $n = 1.225$ so this results in $N = \pi(1.225) = 3.85$. The results for these series of tests are shown in Table 6.7, and it is evident from these images that the random spacing technique is effective.

With the aliasing issue now addressed to some extent, the degrees of freedom demonstration was then run for the targets comprising two circles, one square, two squares, and two triangles as before for the tests with sources. These images are shown in Tables 6.8 through 6.10. As already mentioned, though the aliasing has disappeared, there is now a noise element added to the images from the randomized detector spacing. This being the case, the effect of the degrees of freedom on the reconstructed images is not as pronounced as it was for the demonstrations involving the sources, but there

Table 6.2 Family of First Born Approximation Reconstructions for a Cylinder with a Radius of 60 mm (1λ) Illuminated by a Source with a Frequency of 5 GHz (λ = 60 mm) for a Permittivity Range of 2 to 10

Number of Sources	$\varepsilon_r = 2$ $N = 4.44$ $kVa = 6$	$\varepsilon_r = 3$ $N = 5.44$ $kVa = 13$	$\varepsilon_r = 4$ $N = 6.28$ $kVa = 19$	$\varepsilon_r = 5$ $N = 7.02$ $kVa = 25$	$\varepsilon_r = 6$ $N = 7.70$ $kVa = 31$	$\varepsilon_r = 7$ $N = 8.31$ $kVa = 38$	$\varepsilon_r = 8$ $N = 8.89$ $kVa = 44$	$\varepsilon_r = 9$ $N = 9.42$ $kVa = 50$	$\varepsilon_r = 10$ $N = 9.93$ $kVa = 57$
4									
6									
9									

Note: There are 360 receivers, a resolution of 250 × 250 on each image with increasing number of sources going down the page for each permittivity value. Calculated minimum degrees of freedom (number of sources) and weak scattering metric are shown for reference.

Table 6.3 Family of First Born Approximation Reconstructions for a Cylinder with a Radius of 60 mm (1λ) Illuminated by a Source with a Frequency of 5 GHz ($\lambda = 60$ mm) for a Permittivity Range of 11 to 19

Number of Sources	$\varepsilon_r = 11$ $N = 10.42$ $kVa = 63$	$\varepsilon_r = 12$ $N = 10.88$ $kVa = 69$	$\varepsilon_r = 13$ $N = 11.33$ $kVa = 75$	$\varepsilon_r = 14$ $N = 11.75$ $kVa = 82$	$\varepsilon_r = 15$ $N = 12.17$ $kVa = 88$	$\varepsilon_r = 16$ $N = 12.57$ $kVa = 94$	$\varepsilon_r = 17$ $N = 12.95$ $kVa = 100$	$\varepsilon_r = 18$ $N = 13.33$ $kVa = 113$	$\varepsilon_r = 19$ $N = 13.69$ $kVa = 119$
4									
6									
9									

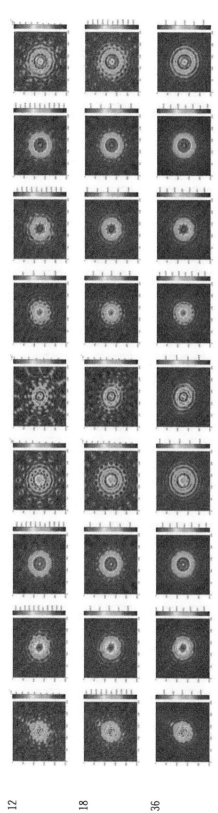

12

18

36

Note: There are 360 receivers, a resolution of 250 × 250 on each image with increasing number of sources going down the page for each permittivity value. Calculated minimum degrees of freedom (number of sources) and weak scattering metric are shown for reference.

Table 6.4 Family of First Born Approximation Reconstructions for a Pair of Cylinders with a Radius of 60 mm (1λ) Illuminated by a Source with a Frequency of 5 GHz (λ = 60 mm) for a Permittivity Range of 2–10

Number of Sources	$\varepsilon_r = 2$ $N = 8.89$ $kVa = 13$	$\varepsilon_r = 3$ $N = 10.88$ $kVa = 25$	$\varepsilon_r = 4$ $N = 12.57$ $kVa = 38$	$\varepsilon_r = 5$ $N = 14.05$ $kVa = 50$	$\varepsilon_r = 6$ $N = 15.36$ $kVa = 63$	$\varepsilon_r = 7$ $N = 16.62$ $kVa = 75$	$\varepsilon_r = 8$ $N = 17.77$ $kVa = 88$	$\varepsilon_r = 9$ $N = 18.85$ $kVa = 101$	$\varepsilon_r = 10$ $N = 19.87$ $kVa = 113$
4									
6									
9									

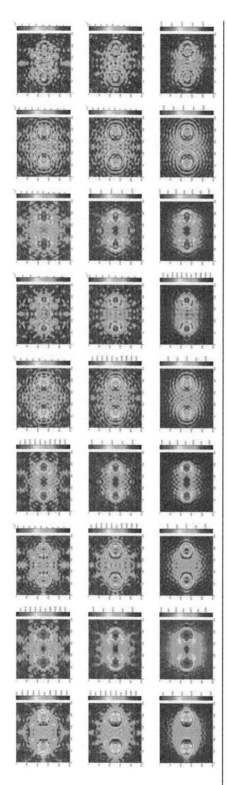

12

18

36

Note: There are 360 receivers, a resolution of 250 × 250 on each image with increasing number of sources going down the page for each permittivity value. Calculated minimum degrees of freedom (number of sources) and weak scattering metric are shown for reference.

Table 6.5 Family of First Born Approximation Reconstructions for a Pair of Squares with Sides of 120 mm (2λ) Illuminated by a Source with a Frequency of 5 GHz ($\lambda = 60$ mm) for a Permittivity Range of 1.1–1.9

Number of Sources	$\varepsilon_r = 1.1$ $N = 8.39$ $kVa = 2.5$	$\varepsilon_r = 1.2$ $N = 8.76$ $kVa = 5$	$\varepsilon_r = 1.3$ $N = 9.12$ $kVa = 7.5$	$\varepsilon_r = 1.4$ $N = 9.47$ $kVa = 10$	$\varepsilon_r = 1.5$ $N = 9.80$ $kVa = 12.6$	$\varepsilon_r = 1.6$ $N = 10.12$ $kVa = 15.1$	$\varepsilon_r = 1.7$ $N = 10.43$ $kVa = 17.6$	$\varepsilon_r = 1.8$ $N = 10.73$ $kVa = 20.1$	$\varepsilon_r = 1.9$ $N = 11.03$ $kVa = 22.6$
4									
6									
9									

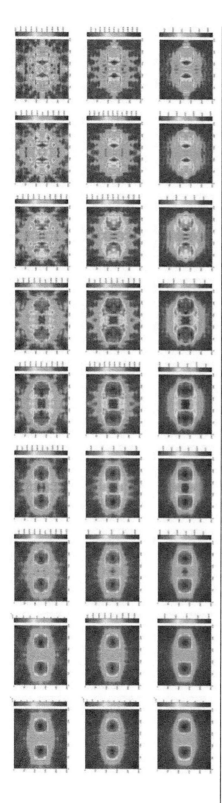

Note: There are 360 receivers, a resolution of 250 × 250 on each image with increasing number of sources going down the page for each permittivity value. Calculated minimum degrees of freedom (number of sources) and weak scattering metric are shown for reference.

Table 6.6 Family of first Born Approximation Reconstructions for a Pair of Triangles with a Base of 120 mm (2λ) and a Height of 180 mm (3λ) Illuminated by a Source with a Frequency of 5 GHz (λ = 60 mm) for a Permittivity Range of 1.1–1.9

Number of Sources	$\varepsilon_r = 1.1$ $N = 6.29$ $kVa = 3.8$	$\varepsilon_r = 1.2$ $N = 6.57$ $kVa = 7.5$	$\varepsilon_r = 1.3$ $N = 6.84$ $kVa = 11.3$	$\varepsilon_r = 1.4$ $N = 7.10$ $kVa = 15.1$	$\varepsilon_r = 1.5$ $N = 7.35$ $kVa = 18.9$	$\varepsilon_r = 1.6$ $N = 7.59$ $kVa = 22.6$	$\varepsilon_r = 1.7$ $N = 7.82$ $kVa = 26.4$	$\varepsilon_r = 1.8$ $N = 8.05$ $kVa = 30.2$	$\varepsilon_r = 1.9$ $N = 8.27$ $kVa = 33.9$
4									
6									
9									

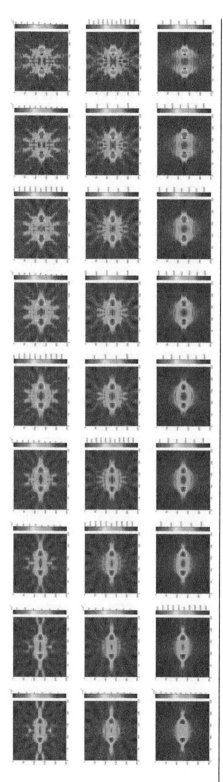

Note: There are 360 receivers, a resolution of 250 × 250 on each image with increasing number of sources going down the page for each permittivity value. Calculated minimum degrees of freedom (number of sources) and weak scattering metric are shown for reference.

is evidence in the images that suggests that the requirement of minimum degrees of freedom on the number of receivers is valid as well. This complements the series of images produced in the previous section.

6.4 IMAGING RELATIONSHIP BETWEEN BORN APPROXIMATION AND MIE Q FACTOR

It is obvious from the data for various cylinders in Tables 6.2 through 6.4 that there appears to be a cyclic pattern to the quality of the Born reconstructed images with increasing scattering strength. As it was demonstrated in the previous two sections that the minimum degrees of freedom have been satisfied, this pattern of behavior must have another explanation. In considering cylindrical targets, it is evident that this circle or cylinder can be analyzed or considered as a 2-D Lorenz–Mie scatterer, as discussed in Section 3.4. This being the case, it would logically follow that, as this cylinder is taken through the different scenarios of varying size and permittivity, the incident wave inside the target should be cycling through various points of resonance. As discussed previously, for Lorenz–Mie scatterers, this is characterized by the following Q factor equation which can be regarded as a measure of the scattering cross section of the cylinder, increasing at resonant frequencies:

$$Q = 2 - \frac{4}{p}\sin p + \frac{4}{p^2}(1 - \cos p) \tag{6.3}$$

where p is defined as

$$p = \frac{4\pi r(n - 1)}{\lambda} \tag{6.4}$$

These two equations are defined almost entirely in terms of physical parameters of the target and the incident wave. These parameters are the radius r, the wavelength of the incident field λ, and the index of refraction n which is also defined as

$$n = \sqrt{\varepsilon_r \mu_r} \tag{6.5}$$

where the relative permeability is generally equal to 1. Tables 6.11 through 6.13 show families of images that demonstrate the effects of Q on Born reconstructions as a function of the permittivity.

From a close examination of the images in these tables, it is quite clear that there does seem to be a cyclic pattern in the quality of the reconstructions as a function of increasing permittivity. Furthermore, examining the graphs in Figures 6.3 through 6.5, it is evident that the performance of the reconstructed images does seem to correlate to the predicted resonances of the target and thus the "good" and "bad" reconstructions can therefore be predicted with some certainty using this information. This then is highly suggestive as a reasonable explanation for why strongly scattering cylinders, near a resonant scattering condition at which the scattering cross section is larger, appear to

Table 6.7 Family of First Born Approximation Reconstructions for a Cylinder with a Radius of 1λ Illuminated by Various Incident Frequencies

Rec#	8 GHz, NBW = 0	8 GHz, NBW = 360/Rec#	5 GHz, NBW = 0	5 GHz, NBW = 360/Rec#	2 GHz, NBW = 0	2 GHz, NBW = 360/Rec#
3						
4						
5						
6						
8						

Note: The target permittivity is 1.5. There are 360 receivers, a resolution of 250 × 250 on each image with increasing number of receivers going down the page. *Green* indicates degrees of freedom satisfied. *Red* indicates that the number of degrees of freedom is insufficient. *Yellow* indicates that the number of available degrees of freedom is marginal. NBW, noise bandwidth.

Table 6.8 Family of First Born Approximation Reconstructions for a Pair of Cylinders with a Radius of 60 mm (1λ) Illuminated by a Source with a Frequency of 5 GHz (λ = 60 mm) for a Permittivity Range of 1.1–1.9

Number of Sources	$\varepsilon_r = 1.1$ $N = 6.59$	$\varepsilon_r = 1.2$ $N = 6.88$	$\varepsilon_r = 1.3$ $N = 7.16$	$\varepsilon_r = 1.4$ $N = 7.43$	$\varepsilon_r = 1.5$ $N = 7.70$	$\varepsilon_r = 1.6$ $N = 7.95$	$\varepsilon_r = 1.7$ $N = 8.19$	$\varepsilon_r = 1.8$ $N = 8.43$	$\varepsilon_r = 1.9$ $N = 8.66$
5									
6									
8									
9									

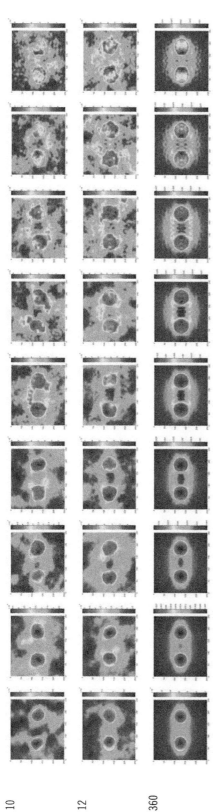

Note: There are 360 sources, a resolution of 250 × 250 on each image with increasing number of receivers going down the page for each permittivity value. Calculated minimum degrees of freedom (number of receivers) are shown for reference.

Table 6.9 Family of First Born Approximation Reconstructions for one Square with a Sides of 120 mm (2λ) Illuminated by a Source with a Frequency of 5 GHz (λ = 60 mm) for a Permittivity Range of 1.1–1.9

Number of Sources	$\varepsilon_r = 1.1$ $N = 4.2$	$\varepsilon_r = 1.2$ $N = 4.38$	$\varepsilon_r = 1.3$ $N = 4.56$	$\varepsilon_r = 1.4$ $N = 4.73$	$\varepsilon_r = 1.5$ $N = 4.90$	$\varepsilon_r = 1.6$ $N = 5.06$	$\varepsilon_r = 1.7$ $N = 5.22$	$\varepsilon_r = 1.8$ $N = 5.37$	$\varepsilon_r = 1.9$ $N = 5.51$
3									
4									
5									
6									

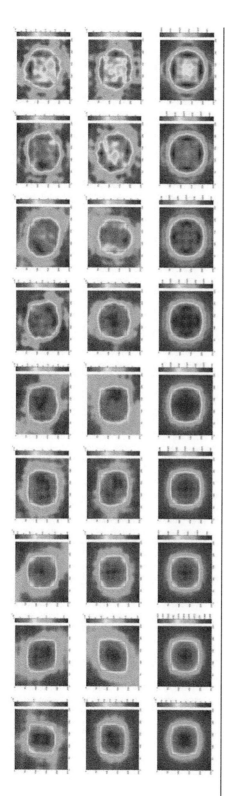

Note: There are 360 sources, a resolution of 250×250 on each image with increasing number of receivers going down the page for each permittivity value. Calculated minimum degrees of freedom (number of receivers) are shown for reference.

Table 6.10   Family of First Born Approximation Reconstructions for a Pair of Squares with Sides of 120 mm (2λ) Illuminated by a Source with a Frequency of 5 GHz (λ = 60 mm) for a Permittivity Range of 1.1–1.9

Number of Sources	$\varepsilon_r = 1.1$ $N = 8.39$	$\varepsilon_r = 1.2$ $N = 8.76$	$\varepsilon_r = 1.3$ $N = 9.12$	$\varepsilon_r = 1.4$ $N = 9.47$	$\varepsilon_r = 1.5$ $N = 9.80$	$\varepsilon_r = 1.6$ $N = 10.12$	$\varepsilon_r = 1.7$ $N = 10.43$	$\varepsilon_r = 1.8$ $N = 10.73$	$\varepsilon_r = 1.9$ $N = 11.03$
6									
8									
9									
10									

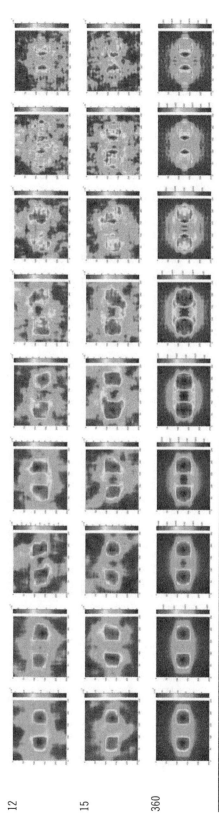

Note: There are 360 sources, a resolution of 250 × 250 on each image with increasing number of receivers going down the page for each permittivity value. Calculated minimum degrees of freedom (number of receivers) are shown for reference.

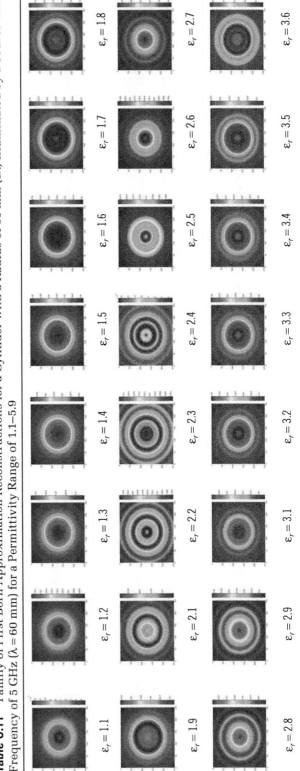

Table 6.11 Family of First Born Approximation Reconstructions for a Cylinder with a Radius of 60 mm (1λ) Illuminated by a Source with a Frequency of 5 GHz ($\lambda = 60$ mm) for a Permittivity Range of 1.1–5.9

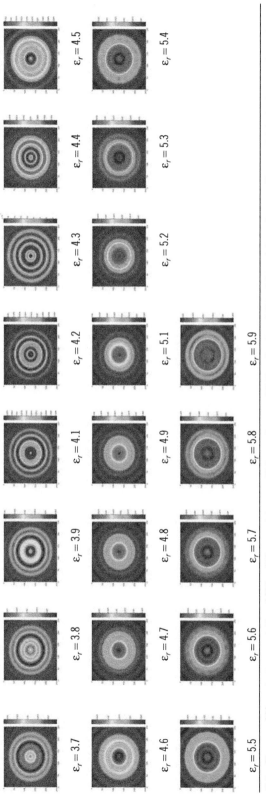

$\varepsilon_r = 4.5$ $\varepsilon_r = 4.4$ $\varepsilon_r = 4.3$ $\varepsilon_r = 4.2$ $\varepsilon_r = 4.1$ $\varepsilon_r = 3.9$ $\varepsilon_r = 3.8$ $\varepsilon_r = 3.7$

$\varepsilon_r = 5.4$ $\varepsilon_r = 5.3$ $\varepsilon_r = 5.2$ $\varepsilon_r = 5.1$ $\varepsilon_r = 4.9$ $\varepsilon_r = 4.8$ $\varepsilon_r = 4.7$ $\varepsilon_r = 4.6$

$\varepsilon_r = 5.9$ $\varepsilon_r = 5.8$ $\varepsilon_r = 5.7$ $\varepsilon_r = 5.6$ $\varepsilon_r = 5.5$

Note: There are 360 receivers, a resolution of 250×250 on each image with increasing permittivity going from left to right and down the page. The permittivity is shown underneath each target.

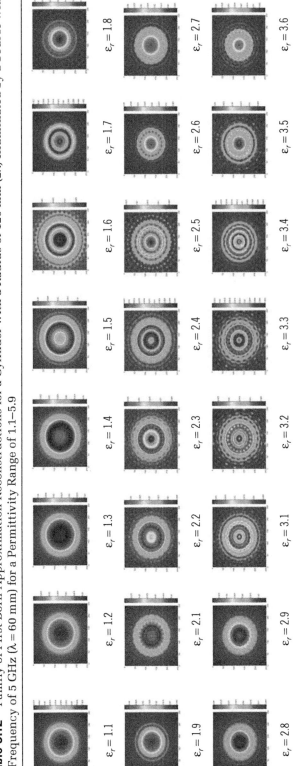

Table 6.12 Family of First Born Approximation Reconstructions for a Cylinder with a Radius of 120 mm (2λ) Illuminated by a Source with a Frequency of 5 GHz (λ = 60 mm) for a Permittivity Range of 1.1–5.9

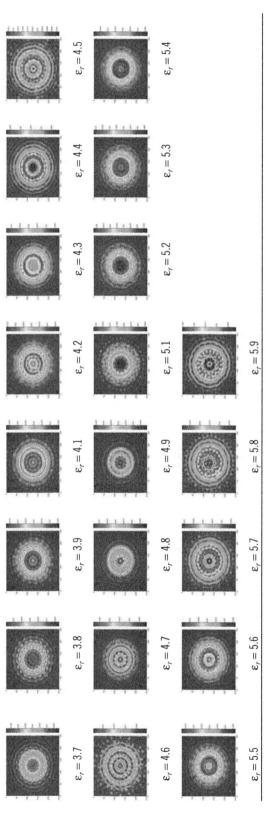

$\varepsilon_r = 3.7$ $\varepsilon_r = 3.8$ $\varepsilon_r = 3.9$ $\varepsilon_r = 4.1$ $\varepsilon_r = 4.2$ $\varepsilon_r = 4.3$ $\varepsilon_r = 4.4$ $\varepsilon_r = 4.5$

$\varepsilon_r = 4.6$ $\varepsilon_r = 4.7$ $\varepsilon_r = 4.8$ $\varepsilon_r = 4.9$ $\varepsilon_r = 5.1$ $\varepsilon_r = 5.2$ $\varepsilon_r = 5.3$ $\varepsilon_r = 5.4$

$\varepsilon_r = 5.5$ $\varepsilon_r = 5.6$ $\varepsilon_r = 5.7$ $\varepsilon_r = 5.8$ $\varepsilon_r = 5.9$

Note: There are 360 receivers, a resolution of 250×250 on each image with increasing permittivity going from left to right and down the page. The permittivity is shown underneath each target.

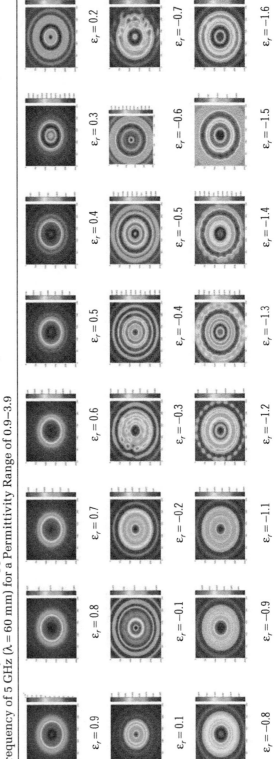

Table 6.13 Family of First Born Approximation Reconstructions for a Cylinder with a Radius of 60 mm (1λ) Illuminated by a Source with a Frequency of 5 GHz (λ = 60 mm) for a Permittivity Range of 0.9–3.9

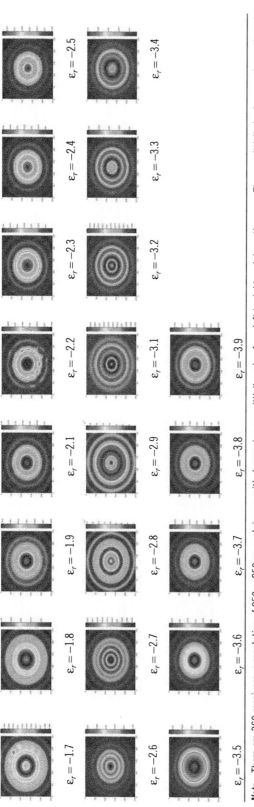

Note: There are 360 receivers, a resolution of 250 × 250 on each image with decreasing permittivity going from left to right and down the page. The permittivity is shown underneath each target. The permeability for each image above is −1.

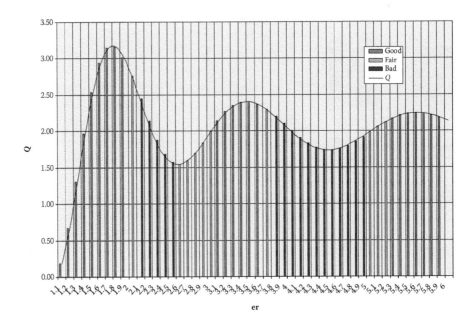

Figure 6.3   Graph showing the plot of the Lorenz–Mie scattering efficiency Q factor for the same conditions as the images in Table 6.11. Superimposed on this graph is a column for each image which indicates the relative quality of the Born reconstruction with green indicating good, yellow indicating fair, and red indicating poor. This graph demonstrates that when Q is "high" or increasing, the image reconstruction is good, and when Q is "low" or decreasing, the image is poor.

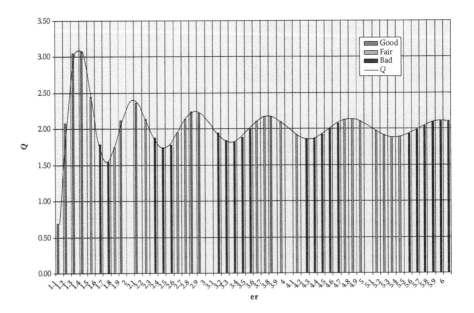

Figure 6.4   Graph showing the plot of the Lorenz–Mie scattering efficiency Q factor equation for the same conditions as the images in Table 6.12. Superimposed on this graph is a column for each image which indicates the relative quality of the Born reconstruction with green indicating good, yellow indicating fair, and red indicating poor. This graph demonstrates that when Q is "high" or increasing, the image reconstruction is good, and when Q is "low" or decreasing, the image is poor.

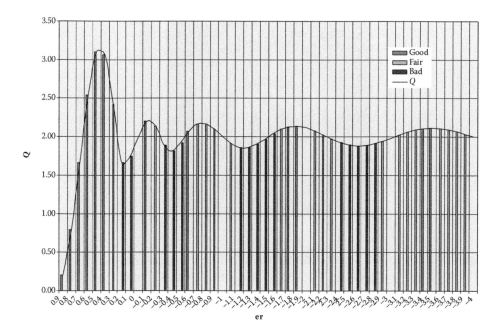

Figure 6.5 Graph showing the plot of the Lorenz–Mie scattering efficiency Q factor equation for the same conditions as the images in Table 6.13. Superimposed on this graph is a column for each image which indicates the relative quality of the Born reconstruction with green indicating good, yellow indicating fair, and red indicating poor. This graph demonstrates that when Q is "high" or increasing, the image reconstruction is good, and when Q is "low" or decreasing, the image is poor.

have a larger than expected area and a more uniform appearance. This also alludes to a new criterion that could be used to help predict the performance of an image reconstruction from scattered fields.

These resonances can be easily predicted for a Mie scatterer, but this may not always be the case for more complex targets. However, resonances for specific target sets could be determined experimentally or via simulation and then possibly characterized by a closed form equation if one uses a structured method developed like the one used here. One could experimentally determine enough resonance points and then use some type of curve-fitting algorithm to determine a resonance relationship much like the Q factor for the Mie scatterers. This is theoretically possible, but for complex structures, a virtual approach like the one used here is more straightforward. Assuming the pattern of resonant frequencies for any given target could be inferred from measurements, then this knowledge could be used to predict the Born approximation-based image reconstruction as well as providing a tool to investigate the properties of resonant scattering structures. One very special case of the Lorenz–Mie model and the scattering cross section indicated by Q is when the permittivity is negative. This is recently of special interest for work currently being done in the meta-material fields as researchers strive to produce materials with negative index of refraction. While this is a challenge in the physical realm, it can be simulated rather easily in the virtual realm. Table 6.13 shows the results for a cylinder as the permittivity is varied from 1 to −4. It should be noted that if the permittivity is allowed to go negative, it causes problems for the Q equation

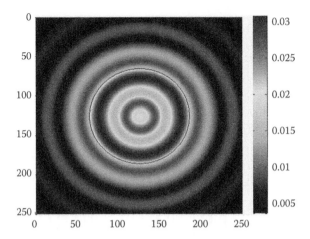

Figure 6.6   The resulting reconstructed image for a 2-D cylinder with a positive value of 1 for the permeability and a varying negative values for the permittivity. All images look virtually the same regardless of the (negative) value for the permittivity.

as p is now an imaginary number and, as is well known, this corresponds to a very lossy or evanescent case of a wave. Thus, when the permittivity is negative and the permeability is positive, the Q relationship breaks down and one would expect no resonances to be predicted. If this condition is simulated using the methods above, the reconstructed images for all of the values in the range of permittivities look like the image in Figure 6.6, which is very poor. To correctly model a negative index material, the permeability value needs to also be set to −1 when the permittivity value is negative. This allows for the negative root to be selected for p which yields a real-valued representation for Q. This gives results consistent with those of the simulation, as seen in Figure 6.5.

REFERENCES

Jones, W. B. 1988. *Introduction to Optical Fiber Communication Systems.* New York: Holt, Rinehart, & Winston, Inc.

Kasap, S. O. 2001. *Optoelectronic Devices and Photonics: Principle and Practices.* New Jersey: Prentice-Hall.

Lustig, M., Donoho, D., and Pauly, J. M. 2010. Sparse MRI: The application of compressed sensing for rapid MR imaging. *Magnetic Resonance In Medicine, 58*(6), 1182–1195.

Miller, D. A. 2007. Fundamental limit for optical components. *Journal of Optical Society of America B, 24*(10), A1–A18.

Ritter, R. S. 2012. *Signal Processing Based Method for Modeling and Solving Inverse Scattering Problems.* University of North Carolina at Charlotte, Electrical Engineering. Charlotte: UMI/ProQuest LLC.

Seven

Alternate Inverse Methods

7.1 ITERATIVE METHODS

Iterative techniques to solve the inverse scattering problem have gained tremendous attention in the past 25 years (Belkebir and Tijhuis, 2001; Chun et al., 2005; Crocco et al., 2005; Dubois et al., 2005; Estataico et al., 2005; Harada et al., 1995; 1990; Isernia et al., 2004; Rieger et al., 1999; Wang and Chew, 1989). Several iterative methods have been proposed but only a few of them have achieved some level of success. We will briefly examine the Born iterative method, distorted Born iterative method, conjugate gradient method, and the prior discrete Fourier transform method which serve as the basis for many other iterative methods and approaches.

7.2 BORN ITERATIVE METHOD (BIM)

Wang and Chew (1989) first proposed the Born iterative method. The first Born approximation is traditionally considered limited for strong scattering objects because of strong diffraction effects; therefore, the inherent nonlinearity of the integral equation in Equation 4.28 has to be taken into account. The starting point of BIM is to first acquire an initial estimate, $V_{BA}^1(\boldsymbol{r})$, of the object by using the first Born approximation. The estimated $V_{BA}^1(\boldsymbol{r})$ is then used to compute the field inside the scattering volume and at the receiver points. The BIM uses a point-matching method with the pulse basis function to solve the forward scattering problem (Wang and Chew, 1989). The estimated field computed in the above step is substituted into Equation 4.7 as follows:

$$\Psi(\boldsymbol{r},\hat{\boldsymbol{r}}_{\text{inc}}) = \Psi_{\text{inc}}(\boldsymbol{r}) - k^2 \int_D V(\boldsymbol{r}')\Psi(\boldsymbol{r}',\hat{\boldsymbol{r}}_{\text{inc}})G_0(\boldsymbol{r},\boldsymbol{r}')\mathrm{d}\boldsymbol{r}' \tag{7.1}$$

to calculate the next order scattering function $V_{BA}^2(\boldsymbol{r})$. The second-order scattering object $V_{BA}^2(\boldsymbol{r})$ is used to solve the scattering problem for the field inside the object and at the observation points. The simulated field Ψ_s^{sim} is then compared with the measured scattered field data Ψ_s^{measured} shown here

$$D = \left| \Psi_s^{\text{sim}}(\boldsymbol{r},\hat{\boldsymbol{r}}_{\text{inc}}) - \Psi_s^{\text{measured}}(\boldsymbol{r},\hat{\boldsymbol{r}}_{\text{inc}}) \right| \tag{7.2}$$

and if the difference between them is less than 5% then the iteration can be terminated, otherwise one continues with the iterations. The BIM also uses a regularization method to address the nonuniqueness and instability of the inverse scattering problem. The regularization method imposes an additional constraint on the linear system and allows the user to choose a solution from many available solutions. Green's function remains unchanged during the entire iterative process. More details of this method can be found in Wang and Chew (1989).

7.3 DISTORTED BORN ITERATIVE METHOD (DBIM)

The distorted Born iterative method was also proposed by Wang and Chew (1990) as an improvement over BIM (Wang and Chew, 1989). Similar to BIM, it starts by solving for the first-order scattering object $V_{BA}^1(r)$ by using the first Born approximation, and the homogeneous Green's function with relative permittivity of unity is used initially. The next step is to use this object function $V_{BA}^1(r)$ to solve the forward scattering problem using the method of moments (Richmond, 1965) and to calculate the field inside the object and at the receiver points. Using $V_{BA}^1(r)$, the point source response in the object for every observation point is computed. In BIM, Green's function was kept constant throughout the iterative process, whereas in DBIM Green's function $G_0^1(r, r')$ is calculated with the last reconstructed permittivity profile as the background permittivity. The estimated Green's function and field are substituted in the integral equation

$$\Psi_s(r, \hat{r}_{\text{inc}}) = -k^2 \int_D V(r')\Psi(r', \hat{r}_{\text{inc}})G_0(r, r')\mathrm{d}r' \qquad (7.3)$$

The calculated scattered field is then subtracted from the field at receivers and the inverse scattering problem is solved for the correction of the last reconstructed profile. Feedback from the previous profile is used to generate a new profile. The forward scattering problem is solved again using this new profile and the computed scattered field is compared with the measured scattered field. If the relative residual error (RRE) is less than the criterion defined then the process terminates; else it continues. Wang and Chew (1989) defined RRE as

$$\text{RRE} = \frac{\sum_{a=1}^{M} \left| \Psi_s^{\text{measured}}(r_a) - \Psi_s^{\text{sim}(j)}(r_a) \right|}{\sum_{a=1}^{M} \left| \Psi_s^{\text{measured}}(r_a) \right|} \qquad (7.4)$$

where j is the iteration cycle. The convergence rate of the distorted Born iterative method is faster than that of the Born iterative method. However, the Born iterative method is more tolerant to noise than the distorted Born iterative method. Depending upon the nature of the problem, either the BIM or DBIM can be utilized.

7.4 CONJUGATE GRADIENT METHOD (CGM)

The conjugate gradient method (Harada et al., 1995; Lobel et al., 1996) is an iterative technique for solving the inverse scattering problem using an optimization procedure. In this method, a functional is defined as a norm of discrepancy between the simulated and the measured scattered field amplitude. Harada et al. (1995) defined this functional as

$$X[V_{est}(\boldsymbol{r})] = \sum_{m=1}^{M} \sum_{n=1}^{N} \left| f\left(V_{est}(\boldsymbol{r}); \phi_{inc}^{(m)}; \phi_{s}^{(n)}\right) - \tilde{f}\left(\phi_{inc}^{(m)}; \phi_{s}^{(n)}\right) \right|^{2} \tag{7.5}$$

where f is the simulated scattered field amplitude which is calculated using an estimated object function $V_{est}(\boldsymbol{r})$, $\tilde{f}$ is the measured scattered field amplitude and $X[V(\boldsymbol{r})]$ is the norm of the difference between the measured and simulated scattered field amplitude. The goal of the optimization method, which in this case is the conjugate gradient, is to find the ideal object function $V_{est}(\boldsymbol{r})$, which minimizes the functional $X[V(\boldsymbol{r})]$. The gradient direction of the functional is found by using the Fréchet derivative (Harada et al., 1995; Lobel et al., 1996; Takenaka et al., 1992). Similar to the Born iterative method, the forward scattering problem is solved using the method of moments (Richmond, 1965) with the pulse basis functions and point matching, which transforms the integral equations into matrix equations. The conjugate gradient method shows good immunity to noise levels, and its convergence rate can be increased by using *a priori* information about the outer boundary of the object.

7.5 PRIOR DISCRETE FOURIER TRANSFORM (PDFT)

The PDFT is a processing step that all methods could potentially benefit from. If we consider function $V\Psi$ in Equation 7.1, in practice, the analyticity of the estimate for $V\Psi$ is assured because of the fact that the data available are always limited in k-space. However, processing this assumes a good estimate of the separable function $V\Psi$, since it is necessary to separate V from Ψ. Since V is assumed to be of finite extent, then $V\Psi$ should also be of compact support; hence, the data in k-space should be analytic. In principle, one may use analytical continuation of the scattered field data in k-space to obtain a better estimate of $V\Psi$ as a product rather than a low pass filtered function and hence a band-limited function. However, it has been observed by several authors (Habashy and Wolf, 1994; Remis and van den Berg, 2000; Roger, 1981) that analytic continuation is not practical because of its instability and sensitivity to noise. The inverse scattering algorithms are Fourier based in nature, and interpolating and extrapolating the Fourier data lying on semicircular arcs in k-space can accomplish improvement in the image quality of the recovered object.

A very stable (regularizable) spectral estimation method known as the PDFT (Byrne et al., 1983; Shieh and Fiddy, 2006), which gives a minimum norm solution of the estimate for $V\Psi$ using properly designed Hilbert spaces, is described here. The success of the PDFT algorithm relies on its flexibility and effective encoding of prior knowledge about the object. The PDFT is a Fourier-based estimator, so it is easy to incorporate it into most signal processing methods. More details of the PDFT algorithm can be found in Shieh and Fiddy (2006), Darling et al. (1983), and the appendices, but in summary, the PDFT assumes that Fourier information about the product of $V\Psi$ is available, which is precisely our measured data in k-space. Let

$$f(\boldsymbol{r}) = V(\boldsymbol{r})\Psi(\boldsymbol{r}) \tag{7.6}$$

then the Fourier transform of $f(\mathbf{r})$ is written as

$$F_n = F(\mathbf{k}_n) = \int\limits_{-\infty}^{\infty}\int\limits_{-\infty}^{\infty} f(\mathbf{r})e^{-i\mathbf{k}_n\mathbf{r}}d^2\mathbf{r} \tag{7.7}$$

for $n = 1, 2, 3, \ldots, N$. The PDFT estimator is given by

$$f_{\text{PDFT}} = p(\mathbf{r})\sum_{m=1}^{M} a_m e^{i\mathbf{k}_m\mathbf{r}} \tag{7.8}$$

where $p(\mathbf{r})$ is the nonnegative prior weighting function containing information about the object. The advantage of the PDFT algorithm is that the data need not to be uniformly sampled, which is why this approach can be used to interpolate and extrapolate both nonuniformly sampled and incomplete data sets. The term PDFT comes from the fact that it is the product of a prior $p(\mathbf{r})$ and the discrete Fourier transform. The coefficients a_m for $m = 1, 2, 3, \ldots, M$ are determined by solving a system of linear equations

$$F_n = \sum_{m=1}^{M} a_m P(\mathbf{k}_n - \mathbf{k}_m) \tag{7.9}$$

In matrix notation, we can write the above equation as

$$\boldsymbol{f} = \boldsymbol{Pa} \tag{7.10}$$

where

$$\boldsymbol{f} = [F(\mathbf{k}_1), F(\mathbf{k}_2), \ldots, F(\mathbf{k}_M)]^T \tag{7.11}$$

$$\boldsymbol{a} = [a_1, a_2, \ldots, a_M]^T \tag{7.12}$$

and where T denotes the transpose of a matrix and $\boldsymbol{P}$ is the $M \times M$ square matrix and is the Fourier transform of prior $p(\mathbf{r})$. Thus, using the coefficients calculated by solving the above set of equations, the PDFT provides a data-consistent image estimate by minimizing the weighted error, which is defined as

$$\xi = \int \frac{1}{p(\mathbf{r})}|f(\mathbf{r}) - f_{\text{PDFT}}(\mathbf{r})|d^2\mathbf{r} \quad \text{(minimum)} \tag{7.13}$$

The above integral is taken over the support of the prior function $p(\mathbf{r})$, which contains the information about the true or estimated support of the object. The computation of the $\boldsymbol{P}$-matrix can be made even if the prior is not available in a closed form but is estimated as a surrounding shape, as the matrix elements can be readily computed. There are also challenges associated with Fourier data being noisy, and a method of regularization usually needs to be employed. If the prior is chosen such that the object energy lies outside the prior support, then without any method of regularization, this will lead to spurious oscillations in the estimate of the object function because the PDFT is data-consistent and requires a lot of energy in the extrapolated spectral values to accommodate an underestimated support. To address this problem, the Tikhonov–Miller regularization is used, which helps in removing the ill-conditioning of the P-matrix by adding a small number to the diagonal elements of the P-matrix (Morris et al., 1997). This is equivalent to adding a small constant amplitude outside the prior support. The regularization step essentially reduces the energy in the estimate arising from the unstable eigenvalues by adding a small value into the prior function (i.e., not allowing it to be zero outside of the support) to slightly increase all eigenvalues without altering the eigenvectors.

If no prior knowledge is available then $p(r)$ would be a constant, and the estimator reduces to the DFT of the available Fourier data. Also, when the regularization constant τ is very large such that $1 + \tau \approx \tau$, the estimator essentially looks like the DFT. The PDFT estimator is both data-consistent and continuous, and works in any number of spatial dimensions version (Duchene et al., 1997). Examples of results using this technique are illustrated later in Chapter 9.

REFERENCES

Belkebir, K. and Tijhuis, A. G. 2001. Modified gradient method and modified Born method for solving a two dimensional inverse scattering problem. *Inverse Problems*, *17*, 1671–1688.

Byrne, C. L., Fitzgerald, R. L., Fiddy, M. A., Hall, T. J., and Darling, A. M. 1983. Image restoration and resolution enhancement. *Journal of the Optical Society of America*, *73*, 1481–1487.

Chun, Y., Song, L. P., and Liu, Q. H. 2005. Inversion of multi-frequency experimental data for imaging complex objects by DTA-CSA method. *Inverse Problems*, *21*, 165–178.

Crocco, L., D'Urso, M., and Isernia, T. 2005. Testing the contrast source extended Born inversion method against real data: The TM case. *Inverse Problems*, *21*(6), S33–S50.

Darling, A. M., Hall, T. J., and Fiddy, M. A. 1983. Stable, noniterative object reconstruction from incomplete data using a priori knowledge. *Journal of the Optical Society of America*, *73*, 1466–1469.

Dubois, A., Belkebir, K., and Saillard, M. 2005. Retrieval of inhomogeneous targets from experimental frequency diversity data. *Inverse Problems*, *21*, 65–79.

Duchene, B., Lesselier, D., and Kleinman, R. E. 1997. Inversion of the 1996 Ipswich data using binary specialization of modified gradient methods. *IEEE Antennas and Propagation Magazine*, *39*(2), 9–12.

Estataico, C., Pastorino, M., and Randazzo, A. 2005. An inexact Newton method for short range microwave imaging within the second order Born approximation. *IEEE Transactions on Geoscience and Remote Sensing*, *43*, 2593–2605.

Habashy, T. and Wolf, E. 1994. Reconstruction of scattering potentials from incomplete data. *Journal of Modern Optics*, *41*, 1679–1685.

Harada, H., Wall, D., Takenaka, T., and Tanaka, M. 1995. Conjugate gradient method applied to inverse scattering problem. *IEEE Transactions on Antennas and Propagation*, *43*(8), 784–792.

Isernia, T., Crocco, L., and D'Urso, M. 2004. New tools and series for forward and inverse scattering problems in lossy media. *IEEE Geoscience and Remote Sensing*, *1*(4), 331–337.

Lobel, P., Kleinman, R. E., Pichot, C., BlancFeraud, L., and Barlaud, M. 1996. Conjugate gradient method for solving inverse scattering with experimental data. *IEEE Antennas and Propagation Magazine*, *38*(3), 48–51.

Morris, J. B., McGahan, R. K., Schmitz, J. L., Wing, R. M., Pommet, D. A., and Fiddy, M. A. 1997. Imaging of strongly scattering targets from real data. *IEEE Antennas and Propagation Magazine*, *39*(2), 22–26.

Remis, R. F. and van den Berg, P. M. 2000. On the equivalence of the Newton-Kantorovich and distorted Born methods. *Inverse Problems*, *16*, L1.

Richmond, J. 1965. Scattering by a dielectric cylinder of arbitrary cross-sectional shape. *IEEE Transactions on Antennas Propagation*, *13*, 334–342.

Rieger, W., Hass, M., Huber, C., Lehner, G., and Rucker, W. M. 1999. Image reconstruction from real scattering data using an iterative scheme with incorporated a priori information. *IEEE Antennas and Propagation Magazine*, *41*(2), 20–36.

Roger, A. 1981. Newton-Kantorovitch algorithm applied to an electromagnetic inverse problem. *IEEE Transactions Antennas and Propagation*, *29*(2), 232–238.

Shieh, H. M. and Fiddy, M. A. 2006. Accuracy of extrapolated data as a function of prior knowledge and regularization. *Applied Optics*, *45*, 3283–3288.

Takenaka, H., Harada, H., and Tanaka, M. 1992. On a simple diffraction tomography technique based on modified Newton-Kantorovich method. *Microwave and Optical Technology Letters*, *5*, 94–97.

Wang, Y. M. and Chew, W. C. 1989. An iterative solution of the two-dimensional electromagnetic inverse scattering problem. *International Journal of Imaging Systems and Technology*, *1*, 100–108.

Wang, Y. M. and Chew, W. C. 1990. Reconstruction of two-dimensional permittivity distribution using the distorted Born iterative method. *IEEE Transactions on Medical Imaging*, *9*, 218–225.

Eight

Homomorphic (Cepstral) Filtering

8.1 CEPSTRAL FILTERING

As previously discussed in Section 4.3, the Born approximation seems to generally perform well when the target is a "weak" scatterer, since one can replace the total field inside the target by the known incident field, and this leads naturally to the Fourier transform relationship described earlier. However, as the permittivity of the target increases, the performance of "Born" algorithm tends to decrease. This is not unexpected due to the fact that less of the incident wave might be penetrating and propagating through the target, but is reflected off the surface of the target as well as being scattered multiple times from inhomogeneities that exist inside the target. The reflected wave emerging from more highly structured scattered field components needs to be interpreted as carrying information about $V(r)$ (see Equations 4.7 and 4.28) or, depending on how the scattered field is processed, as noise-like terms arising from strong scattering that one might be able to remove. In the latter case, if the noise-like terms can be identified and removed or attenuated, then this can ideally reduce a strongly scattering target to one that can be imaged more like a weakly scattering one. In the cepstral method, the total field estimated within the target volume is regarded as a form of spatial noise to be removed. When this is justified, the reconstructed image of the target based on assuming the first Born approximation can be expressed as

$$V_{BA}(r, \hat{r}_{inc}) \approx V(r) \frac{\Psi(r, \hat{r}_{inc})}{\Psi_{inc}(r, \hat{r}_{inc})} \tag{8.1}$$

where $\langle \Psi(r, \hat{r}_{inc})/\Psi_{inc}(r, \hat{r}_{inc}) \rangle$ is a symbolic representation for a complex and noise-like term with a characteristic range of spatial frequencies dominated by the average local wavelength of the source. The problem is now reduced to a complex filtering problem in which the multiplicative term $\langle \Psi(r, \hat{r}_{inc})/\Psi_{inc}(r, \hat{r}_{inc}) \rangle$ needs to be filtered out of the data.

There are a number of ways to approach this problem, but one of the more recent methods is a well-known and documented method based on cepstral filtering, a technique originally developed to eliminate multiplicative noise (Childers et al., 1977; Raghuramireddy and Unbehauen, 1985). The homomorphic filtering method uses the log operation to convert a multiplicative modulation relationship into an additive relationship. This is demonstrated here by taking the complex logarithm of Equation 8.1 as follows:

$$\log\left(V(\mathbf{r})\left\langle\frac{\Psi(\mathbf{r},\hat{\mathbf{r}}_{\text{inc}})}{\Psi_{\text{inc}}(\mathbf{r},\hat{\mathbf{r}}_{\text{inc}})}\right\rangle\right) = \log|V(\mathbf{r})| + \log\left|\left\langle\frac{\Psi(\mathbf{r},\hat{\mathbf{r}}_{\text{inc}})}{\Psi_{\text{inc}}(\mathbf{r},\hat{\mathbf{r}}_{\text{inc}})}\right\rangle\right|$$
$$+ i\left[\arg[V(\mathbf{r})] + \arg\left(\left\langle\frac{\Psi(\mathbf{r},\hat{\mathbf{r}}_{\text{inc}})}{\Psi_{\text{inc}}(\mathbf{r},\hat{\mathbf{r}}_{\text{inc}})}\right\rangle\right)\right]$$

(8.2)

If the Fourier transform is now taken of Equation 8.2, the results will be the complex cepstrum representation of the target or $V(\mathbf{r})\Psi/\Psi_{\text{inc}}$ (Childers et al., 1977; Raghuramireddy and Unbehauen, 1985). With the data now in this form, the phase information of the complex data is retained which is essential in doing any type of useful filtering processes. It should be noted that there are numerous potential problems that can occur when calculating the complex $\log(V\Psi)$ and taking its Fourier transform. There can be unwanted harmonics introduced by taking the complex logarithm of sampled data where their magnitude is less than 1 and approaches zero, which could lead to aliasing in the cepstral domain (Childers et al., 1977). There is also the common concern of branch cuts associated with singularities caused by zeros in the data representation of $V(\mathbf{r})\Psi$ since the logarithm function of zero is strictly undefined. Thus, the complex logarithmic function can be multivalued if the imaginary part of the data representing $V(\mathbf{r})\Psi$ exceeds 2π. When this happens, this results in what is called phase wrapping, which can be extremely difficult to deal with, especially in two or more dimensions. These phase wrappings present discontinuities which generate many frequencies in the Fourier domain, being associated with zeros in the field data corresponding to dislocated wave fronts (Fiddy and Shahid, 2003).

8.2 CEPSTRAL FILTERING WITH MINIMUM PHASE

In previous research (Shahid, 2009), it has been demonstrated that if the cepstral data are made to be causal and minimum phase, and a spatial filter in the cepstral domain is chosen properly, the ability to eliminate, or at least greatly reduce or attenuate, the presence of the $\langle\Psi(\mathbf{r},\hat{\mathbf{r}}_{\text{inc}})/\Psi_{\text{inc}}(\mathbf{r},\hat{\mathbf{r}}_{\text{inc}})\rangle$ term can be achieved leading to a much better reconstructed image of $V(\mathbf{r})$ than that obtained using the Born approximation alone. The first two characteristics mentioned above, that is, forcing the data to be causal and to be minimum phase, are crucial in the success of this approach. These conditions are imperative as demonstrated in Shahid (2009) and they form the basis for the success of this approach. This approach will be thoroughly examined and extended in this book; so this being the case, it seems prudent to review the aspects of this approach here.

The issue of processing the data to be causal is a simple one in that this can be achieved by shifting all of the data into the first quadrant of an expanded space whose origin is defined to be at its center. This is basically reduced to a function of "re-indexing" the indices of the data in the computer code. The second aspect mentioned above is that of making the data to be minimum phase. In 1-D this concept is well understood as discussed in Dudgeon and Mersereau (1984). In this case, a 1-D signal, $f(x)$, is said to be minimum phase if and only if its Fourier transform, $F(z)$, has a zero-free upper-half plane. In other words, $F(z)$ has no zeros for $v > 0$ where $z = u + iv$ and

$$F(u) = \int\limits_{-\infty}^{+\infty} f(x)e^{i2\pi ux}\mathrm{d}x \qquad (8.3)$$

Functions of this type have some very useful properties (Dudgeon and Mersereau, 1984), such as

1. $f(x)$ is causal.
2. The phase of $F(u)$ lies between $-\pi$ and $+\pi$; that is, its phase is always unwrapped (nondiscontinuous).
3. Most of the energy in $f(x)$ lies close to origin.
4. $F(u)$ is absolutely summable.

Another very interesting and important aspect of minimum phase functions is that these functions have a "minimum energy delay" property. This is important because it results in a function that possesses the highest "partial" energy among all functions that have the same Fourier magnitude (Oppenheim et al., 1999). As stated before, a function, $f(x)$, is defined as minimum phase if its Fourier transform, $F(u)$, is analytic and has a zero-free half plane.

With these definitions in hand, the concept of complex cepstrum (Bogert et al., 1963) will now be defined and discussed. The complex cepstrum for a function f is defined as taking the complex natural logarithm of the Fourier transform of the function f, then taking the inverse Fourier transform. This can be written mathematically for signal processing applications utilizing the FFT as

$$\hat{g} = ifft\big[\log(F)\big] \quad \text{where} \quad F = fft(f) \qquad (8.4)$$

or, another common way of expressing this is as follows

$$\hat{g}(x) = \frac{1}{2\pi} \int\limits_{-\infty}^{+\infty} \log\big[F(u)\big]e^{i2\pi ux}\mathrm{d}u \qquad (8.5)$$

where $\hat{g}$ is defined as the cepstrum of the input signal f. As alluded to earlier, in a 1-D case, the minimum phase condition is required to ensure that the Fourier transform of a causal function has a zero-free upper-half plane. This is important because this now permits Cauchy's integral formula to be used to relate the Fourier magnitude and the phase on the real u-axis. This relationship is known as the logarithmic Hilbert transform.

As discussed in Shahid (2009), the concept of zero-free half plane in 2-D can be problematic, to say the least. Even for a function that is separable such as $F(u_1, u_2) = G(u_1)H(u_2)$, a zero-free upper-half plane in $G(z_1)$ automatically leads to $F(z_1, z_2)$ having zeros in the upper-half plane of z_2, since we know that $H(u_2)$ must have zeros. In practice, functions having a zero-free half plane are rare, and there are few general conditions that are known for which these characteristics can be imposed.

In general, in terms of Fourier-based theory and analysis of scattering and inverse scattering, scattered and propagating fields are analytic functions due to the fact that the scattering objects themselves are of finite spatial extent. To go one step further, these fields are entire functions of the exponential type (Burge et al., 1976), which by definition means that they satisfy the Cauchy–Riemann equation for all finite $z = x + iy$ as expressed here:

$$\frac{\partial \mathrm{Re}F}{\partial x} = \frac{\partial \mathrm{Im}F}{\partial y} \tag{8.6}$$

$$\frac{\partial \mathrm{Re}F}{\partial y} = \frac{-\partial \mathrm{Im}F}{\partial x} \tag{8.7}$$

There are of course very strict constraints on these types of functions as to how their amplitude varies or increases and how their zero crossings are distributed in general. In 1-D this is well understood and these concepts are, for the most part, indirectly applicable in the 2-D realm.

8.3 GENERATING THE MINIMUM PHASE FUNCTION

Dudgeon and Mersereau (1984) state that a 2-D minimum phase function is one that is absolutely summable and the inverse and complex cepstrum of which are also absolutely summable and have the same region of support. This region of support also has to be a convex region of some kind. It has not been possible to find general properties for classes of functions for which these conditions can be satisfied, and this condition appears to be sufficient but not necessary. Some specific examples exist of well-conditioned cepstra, as a result of the incorporation of a background or reference wave on the function whose logarithm is to be Fourier transformed. Taking the logarithm of a band-limited function produces a band-limited function and hence a cepstrum of finite support only for minimum phase functions.

At this point, a brief digression on dispersion relations may be helpful. It is well known that the real and imaginary parts of the Fourier transform of a causal function $f(x)$ are related to each other by a Hilbert transform. This follows directly from Titchmarsh's theorem (Fiddy and Shahid, 2003) and is a consequence of the finite support or causal nature of the Fourier transform. The Hilbert transform, HT, is an integral transform that is a principal value integral, which is implicitly solved as a contour integral. The contour is the real axis and a semicircle in either the upper or lower half of the complex plane, which, if Jordan's lemma is satisfied, can be ignored. The HT relationship then takes the form

$$\mathrm{Im}\big[F(u)\big] = \frac{1}{\pi} P \int_{-\infty}^{+\infty} \frac{\mathrm{Re}[F(u')]}{u' - u} \mathrm{d}u' \tag{8.8}$$

$$\mathrm{Re}\big[F(u)\big] = -\frac{1}{\pi} P \int_{-\infty}^{+\infty} \frac{\mathrm{Im}[F(u')]}{u' - u} \mathrm{d}u' \tag{8.9}$$

where P is the Cauchy principal value. Closure of the contour without contributions to the value of the integral from the residues is necessary and so the real and imaginary parts of $\log[F(u)]$ can be similarly related, provided F has no zeros within the contour of integration. This statement also defines the minimum phase condition. Function F can be written in terms of magnitude and phase as

$$F = |F|e^{i\varphi} \tag{8.10}$$

Taking the complex logarithm of (8.10) yields

$$\log(F) = \log|F| + i\varphi \tag{8.11}$$

where $\log|F|$ is the real part and φ is the imaginary part. One can compute the phase of F

$$\text{Im}\left[\log F(u)\right] = \frac{1}{\pi} P \int_{-\infty}^{+\infty} \frac{\text{Re}[\log F(u')]}{u' - u} du' \tag{8.12}$$

The above integral only works if F has no zeros in the upper-half plane. If F is not a minimum phase function, then the result of the above integral will be to produce a phase that when applied to $|F|$ generates a minimum phase function. An important feature of a minimum phase function is that the phase is a continuous function bounded between $-\pi$ and π, and "minimum" in this sense can be interpreted to mean that the phase is already unwrapped. Using this property of minimum phase functions, we can execute the step defined in Equation (8.2) in a satisfactory fashion.

In the 1-D problem, it is possible to enforce the minimum phase condition on a function by applying Rouche's theorem (Fiddy, 1987). A 2-D version of Rouche's theorem has been validated in Shahid et al. (2005). Suppose $h = F(z)$ is analytic in a domain D where $F = (F_1, F_2,..., F_N)$ and the boundary of D is smooth and contains no zeros of F, then if for each point z on the boundary, there is at least one index j $(j = 1, 2,..., N)$ such that $|F_j(z)| > |G_j(z)|$ then $G(z) + F(z)$ have the same number of zeros in D as the number of zeros in $F(z)$ (it actually suffices that $\text{Re}\{G_j(z)\} < \text{Re}\{F_j(z)\}$). In other words, if a function A has N number of zeros; B has M number of zeros, and $|A| > |B|$ on same contour, then $A + B$ will have N number of zeros in that contour contrary for $|A| < |B|$, then $A + B$ will have M number of zeros (see Figure 8.1). The sum of the two functions will have the number of zeros equal to the number of zeros of the larger magnitude function. Consequently, adding a sufficiently large background or reference wave to a band-limited function A, where we

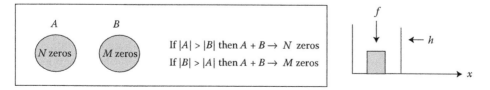

Figure 8.1 Pictorial description of Rouche's theorem.

assume $|A| \ll 1$, allows us to write $G = 1 + A \sim \exp(A)$, thereby satisfying this minimum phase condition.

Therefore, if to our band-limited function $F(z)$ we add another function $G(z)$ which we refer to as a reference function and $G(z)$ has no zeros in upper-half plane, then the function $F(z) + G(z)$ will have no zeros in upper-half plane, thus satisfying Rouche's theorem and the minimum phase condition. The addition of a reference function to an analytic function only moves the zeros from upper-half plane to lower-half plane without destroying them (Fiddy, 1987). If a reference function is chosen as a constant, then we can always find a contour in the upper-half plane along which the magnitude of the added function is greater than the magnitude of $F(z)$. Increasing the constant moves the contour across the real axis and thus pushing zeros to the lower-half plane. It therefore follows that one can preprocess by adding a finite constant or reference point to make it minimum phase before taking its logarithm.

8.4 PREPROCESSING DATA

The preprocessing step requires the data available in k-space to be made causal, $V\langle\Psi\rangle_c$. This can be done by moving available scattered field data into one quadrant of k-space, that is, data are nonzero only in one quadrant.

The next step is to add a reference point at the origin of the causal data $V\langle\Psi\rangle_c$ in k-space. This corresponds to adding a linear phase factor to $V\Psi$ in the object domain. In order to satisfy the minimum phase condition the reference point R does not need to have an arbitrarily large amplitude (a sufficient condition), but simply be just large enough to ensure that phase of $V\Psi$ is continuous and lies within the bounds of $-\pi$ and $+\pi$. A very large reference point with amplitude R such that $|V\Psi|/R \ll 1$ leads to the Fourier transform of $\log(R + V\langle\Psi\rangle) \rightarrow \log(1 + V\langle\Psi\rangle/R)$ being approximately equal to $V\langle\Psi\rangle/R$ indicating that we have satisfied the minimum phase condition but we have not provided a function of $V\langle\Psi\rangle$ that will result in the successful filtering in the cepstral domain.

In this case, the cepstrum of $V\langle\Psi\rangle$ contains the same information that we originally had in k-space. This corresponds to $R = 0$ and the unwanted harmonics in the cepstrum makes filtering impossible. The optimal choice of R is an amplitude which is just large enough to ensure that the phase of $V\langle\Psi\rangle$ is unwrapped and lies between $-\pi$ and $+\pi$. It has been shown (Fiddy and Shahid, 2003; McGahan and Kleinman, 1997) that in order to enforce a minimum phase condition, the reference should satisfy

$$R \geq |V\langle\Psi\rangle|_{max} \tag{8.13}$$

that is, the reference point needs to be larger than the maximum value of the scattered field amplitude to satisfy Rouche's theorem. It is evident that the inequality given in Equation (8.13) is a sufficient condition to enforce minimum phase condition. In the forward scattering problem, this is intuitively satisfying since it is equivalent to requiring that the scattering is relatively weak compared to a larger amplitude background or incident plane wave, as demanded by the first Born approximation. The analogy with holography is also immediate since interference with a reference wave, especially when off-axis with respect to the scattered field, ensures that the phase of the scattered field is encoded in the magnitude via a logarithmic Hilbert transform. An important point, however, is that if one can compute a good estimate for $V\langle\Psi\rangle$ from measured data, then one could numerically add in this reference wave prior to homomorphic filtering. It becomes a postdata processing step and not an experimental requirement.

In order to choose an appropriate reference point, the starting point is to make the amplitude of the reference point the same as the maximum amplitude of $V\langle\Psi\rangle$ and then systematically increase its amplitude until the phase of Born reconstructed $V\langle\Psi\rangle$ lies between $-\pi$ and $+\pi$. Figure 8.2 shows the result of this procedure applied to a cylinder of radius 1λ.

This value for R permits us to write

$$\log(R + V\langle\Psi\rangle) \to \log\left(1 + \frac{V\langle\Psi\rangle}{R}\right) : \frac{V\langle\Psi\rangle}{R} - \frac{1}{2}\left(\frac{V\langle\Psi\rangle}{R}\right)^2 + \cdots \tag{8.14}$$

or one could also write

$$\log\left(1 + \frac{V\langle\Psi\rangle}{R}\right) = \log\left(\left\{\frac{V}{R}\right\}\left\{\frac{R}{V} + \langle\Psi\rangle\right\}\right) = \log\left(\frac{V}{R}\right) + \log\left(\frac{R}{V} + \langle\Psi\rangle\right) \tag{8.15}$$

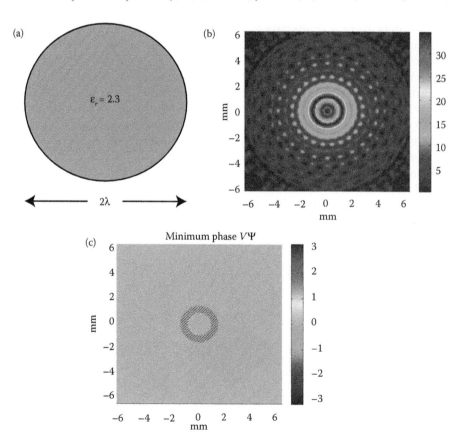

Figure 8.2 Scattering cylinder. (a) Cylinder with 2λ diameter and relative permittivity of 2.3, (b) Born reconstruction, and (c) minimum phase after adding a suitable reference point.

$$\approx \log\left(\frac{V}{R}\right) + \log\left(R' + \langle\Psi\rangle\right)$$

$$\approx \log\left(\frac{V}{R}\right) + \frac{\langle\Psi\rangle}{R'} \tag{8.16}$$

The second term in Equation (8.16) contains spatial frequencies, which are similar to the spatial frequencies of the incident field. One can vary the frequency of the incident plane wave to determine the spatial frequency characteristics of the second term in Equation (8.16) whereas log(V/R) should stay the same. The implementation of the cepstral filtering requires that a low pass filter be applied until the wavelike structure associated with $V\langle\Psi\rangle$ is removed. The success of the filtering operation depends upon how distinct the spatial frequencies of $V\langle\Psi\rangle/R'$ are from spatial frequencies of log(V/R). A linear combination of estimates for V acquired in this way will further improve the signal-to-noise ratio (SNR) of the reconstructed V while suppressing any residual components from the Fourier transform in each of these images.

8.5 TWO-DIMENSIONAL FILTERING METHODS

Once the data from a scattering image has been properly processed as discussed in the previous section, and mapped into the cepstrum domain, it is now necessary to filter the data in an attempt to eliminate or at least attenuate the Ψ "noise" term and obtain a better representation of the original target, $V(\mathbf{r})$. Since the preprocessed data is now minimum phase as demonstrated in the previous section, we know, as stated earlier, that one of the main characteristics of minimum phase data is that most of the energy related to $V(\mathbf{r})\langle\Psi(\mathbf{r})\rangle$ should be located near the origin. This would suggest that some type of low pass filter should work well if chosen properly. In basic filter theory (Gonzalez and Woods, 1992; Jackson, 1991), there are two basic types of low pass filters to consider. They are the ideal low pass filter illustrated in Figure 8.3, and the Gaussian low pass filter illustrated in Figure 8.4. In each of these figures, a top view, isometric view, and slice or profile view is shown for each filter.

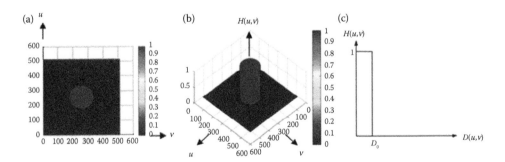

Figure 8.3 Low pass filter. (a) Ideal hard-cut low pass filter 2-D view, (b) low pass filter displayed in 3-D, and (c) the cross section of ideal low pass filter.

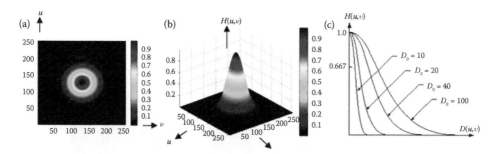

Figure 8.4 Gaussian low pass filter. (a) Gaussian low pass filter 2-D view, (b) Gaussian low pass filter displayed in 3-D, and (c) cross section of Gaussian low pass filter.

In addition, for the Gaussian filter, the profile is shown for various values of sigma, which in effect defines the filter bandwidth.

As is commonly known in basic filter theory (Jackson, 1991), and verified in Shahid (2009), the performance of the Gaussian is generally better than the ideal filter since the ideal filter is prone to produce "ringing" in the reconstructed image as illustrated in Figure 8.5. This being the case, the Gaussian filter is the preferred filtering process in the cepstrum domain. This still leaves the question as to what are the optimum values for the parameters for this filter. More specifically, what should the value of sigma be for the filter, which governs the bandwidth of the filter, and what are the constraints and effects for the peak value of the filter? These parameters will be examined in detail in the following section.

8.6 REMOVING THE REFERENCE

One final technique that will be demonstrated in this chapter pertaining to filtering or attenuating the Ψ term is that of subtracting a cepstrum version of illuminating incident wave Ψ_0 in the cepstrum domain in an attempt to improve the reconstructed image, or possibly improve the scale of the reconstructed image. If we review for a moment the final version of the $V\Psi$ product, which strictly corresponds to $V\Psi/\Psi_{\text{inc}}$, obtained in Equation (8.16), we have that this product can be approximated as

$$\approx \log\left(\frac{V}{R}\right) + \frac{\langle\Psi\rangle}{R'}$$

As stated earlier it is evident that there is a "noise" term in this expression that appears to be of the form of a "weighted" incident frequency type term, depending on how strong a scatterer the target is. This being the case, it seems that it would be beneficial to take the incident field and obtain a weighted cepstrum representation of it and subtract it from the expression above. Assuming that the correct weighting factor is found, it appears that this would have some positive benefit to the resulting cepstrum representation of $V(\mathbf{r})$. Moreover, there will be some benefit to trying to eliminate or account for the reference "R" term as well that was inserted to satisfy Rouche's theorem. This obviously should have some effect on the overall scaling of the

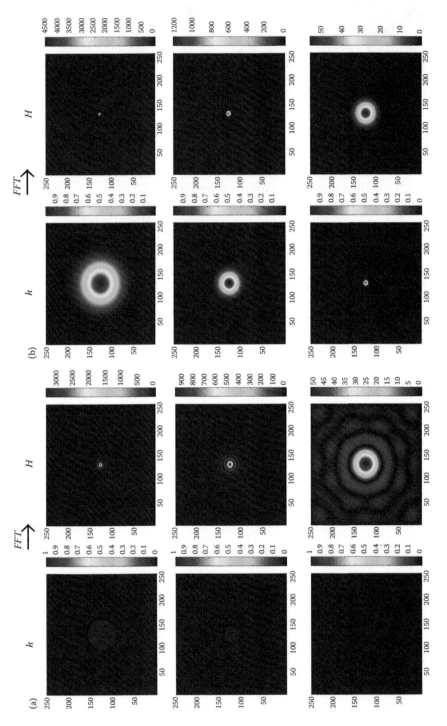

Figure 8.5　Hard-cut filter versus smoothed filter—effect of ringing. (a) Ideal 2-D circular hard-cut filter and its frequency spectrum and (b) 2-D Gaussian low pass filter and its frequency spectrum.

resulting data in theory. If a close representation for the weighted Ψ term can be found, then the resulting form of the expression above would be

$$\approx \log\left(\frac{V}{R}\right) + \frac{\langle\Psi\rangle}{R'} - \kappa\langle\Psi_{\text{inc}}\rangle \qquad (8.17)$$

Again, if the factor for "κ" can be found that is close to R' and $\langle\Psi_{\text{inc}}\rangle$ is a close approximation to $\langle\Psi\rangle$, then improvement in the resulting reconstructed image can be expected. This too will be demonstrated in the following chapters.

REFERENCES

Bogert, B. P., Healy, M. J., and Tukey, J. W. 1963. The frequency analysis of time series echoes: Cepstrum, psuedo-autocovarience, cross-cepstrum and saphe cracking. In M. Rosenblat (Ed.), *Proceedings of the Symposium on Time Series Analysis* (pp. 209–243). New York: Wiley.

Burge, R. E., Fiddy, M. A., Greenaway, A. H., and Ross, G. 1976. The phase problem. In *Proceedings of the Royal Society, A350*, 191–212. London.

Childers, D. G. Skinner, D. P., and Kemarait, R. C. 1977. The ceptstrum: A guide to processing. *IEEE Procedures, 65*(10), 1428–1443.

Dudgeon, D. E. and Mersereau, R. M. 1984. *Multidimensional Digital Signal Processing.* New Jersey: Prentice-Hall.

Fiddy, M. A. 1987. The role of analyticity in image recovery. In H. Stark (Ed.), *Image Recovery: Theory and Applications* (pp. 499–529). Florida: Academic Press.

Fiddy, M. A. and Shahid, U. 2003. Minimum phase and zero distributions in 2D. *Proceedings of SPIE, 5202*, 201–208.

Gonzalez, R. C. and Woods, R. E. 1992. *Digital Image Processing.* New York: Addison-Wesley.

Jackson, L. B. 1991. *Digital Filters and Signal Processing* (2nd ed.). Norwell, MA: Kluwer Academic Publishers.

McGahan, R. V. and Kleinman, R. E. 1997. Second annual special session on image reconstruction using real data. *IEEE Antenna Propagation Magazine, 39*(2), 7–32.

Oppenheim, A. V., Schafer, R. W., and Buck, J. R. 1999. *Discrete-time signal processing.* New Jersey: Prentice-Hall.

Raghuramireddy, D. and Unbehauen, R. 1985. The two-dimensional differential cepstrum. *IEEE Transactions on Acoustics, Speech and Signal Processing, 33*(5), 1335–1337.

Shahid, U. 2009. *Signal Processing Based Method for Solving Inverse Scattering Problems.* PhD Dissertation, Optics. University of North Carolina at Charlotte, Charlotte: UMI/ProQuest LLC.

Shahid, U., Testorf, M., and Fiddy, M. A. 2005. Minimum-phase-based inverse scattering algorithm. *Inverse Problems, 21*, 1–13.

III

APPLICATIONS

Nine

Applications to Real Measured Data

9.1 IPSWICH DATA RESULTS

In this section, various methods are applied to real data provided by various groups. The US Air Force Research Laboratory (AFRL) initiated the idea that it would be beneficial for the inverse scattering community to test their algorithms on measured data from unknown targets. They conducted a number of scattering experiments and provided scattered field data, known as Ipswich data, in the mid-1990s (McGahan and Kleinman, 1999) and which has still kept the community busy since then. More recent data sets have been the focus of special issues of journals such as inverse problems using data provided by the Institut Fresnel (McGahan and Kleinman, 1999). In addition, the proposed methods in this book are applied on the analytic data to study the effect of sampling on the inverse scattering problem.

The AFRL collected Ipswich data in an anechoic chamber, using the swept-bistatic system described in Maponi et al. (1997). Figure 9.1 taken from McGahan and Kleinman (1999) shows the layout of the measurement system, and defines the angles used to describe the data.

9.1.1 IPS008

First we perform the minimum phase procedure on Ipswich data sets IPS008 and IPS0010, the strong scatterers. The IPS008 test subject consists of two cylinders; the outer cylinder is filled with sand and the inner cylinder is filled with salt. The cylinder is placed far enough from the source to ensure that the illuminating wave is well approximated by a plane wave.

The measured and limited data consisted of 36 illumination directions, at equal angular separations of 10° and 180 complex scattered field measurements for each view angle using a frequency of 10 GHz. These data were located on arcs in k-space and moved into one quadrant to impose causality. The IPS008 object represents a strongly scattering penetrable object and has proved to be one of the most difficult to recover from its scattered field data (McGahan and Kleinman, 1999). The geometry of IPS008 is shown in Figure 9.2a; the best image one could expect from the available scattered field measurements, assuming only an inverse Fourier transform is necessary over the loci of data circles in k-space, is shown in Figure 9.2b.

Figure 9.2c shows a reconstruction from the nonlinear filtering method. The reference point, R, introduced at the origin in k-space was increased until the phase of $V(r)\langle\Psi(r,\hat{r})\rangle$ lies between $+\pi$ and $-\pi$. A Gaussian low pass filter was then applied until all wave-like features had been eliminated from the

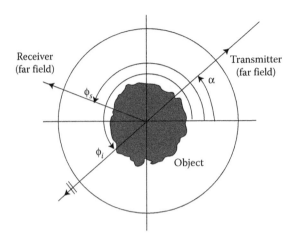

Figure 9.1   Ipswich scattering experiment where α is the angle of incidence, φ_s is the scattering angle, and φ_i the illuminating direction.

resulting reconstruction of the object, V. The IPS008 target is considered a strong scattering object since $k|V|a \approx 87$ where k, the wave number, is calculated as $k = 2\pi/\lambda = 2\pi/0.03 = 209.5$, V is the scattering strength or average permittivity and a is the dimension of the largest feature of object. IPS008 has proved to be a challenging object to image for all of the inverse scattering

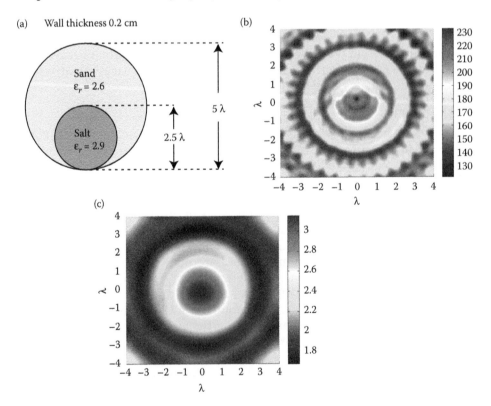

Figure 9.2   IPS008 target configuration: (a) Target object IPS008 geometry, (b) Born reconstruction, (c) cepstral reconstruction, and (d) geometrical comparison between reconstructed object and original object.

community as most of the algorithms applied are typically based on the first Born approximation, and clearly the image obtained using the Born approximation shows poor reconstruction (Figure 9.2b). When an incident plane wave impinges on IPS008, it gets diffracted at the boundary of the larger cylinder. As a result, the wave that is incident on the smaller cylinder can be interpreted as a convergent wave front. This constitutes a serious violation of the first-order Born approximation, which assumes the incident field to pass the target essentially unperturbed, as a result of which we see artifacts in the reconstruction as shown in Figure 9.2b. Figure 9.2c shows the reconstruction obtained using the cepstral method discussed earlier in this book. The image of Figure 9.2c looks much better as compared with an estimate using the Born approximation. The step-by-step process of recovering a scattering object using cepstral filtering method is shown in Figure 9.3. It not only gives an improved estimate of the shape of object but also provides a good approximation of the permittivity distribution. The cepstral filtering method does a good job in recovering the dimensions of cylinders; however, the recovered inner cylinder is not tangent to the external one as it should be.

It is suspected that this difference in the position of the inner cylinder is due to the limited data coverage. It will be shown that the spectral estimation technique, prior discrete Fourier transform (PDFT), can help in resolving this position offset.

Figure 9.4 shows the comparison of reconstructed IPS008 object from various inverse scattering groups published in *IEEE Antennas and Propagation Magazine* (Belkebir and Saillard, 2001; Byrne and Fitzgerald, 1984; Estatico et al., 2005) and a minimum phase-based reconstruction.

Further improvements could possibly be realized by applying the PDFT techniques described in Chapter 7. An example of the effects of this technique in conjunction with cepstral filtering and apparent improvements in terms of dimensions for the IPS008 data is illustrated in Figure 9.5.

9.1.2 IPS010

The IPS010 target consists of a dielectric wedge made of Plexiglass with relative permittivity $\varepsilon_r \approx 2.25$. The IPS010 has also proven to be a very challenging object for participating groups to image due to its high permittivity and sharp shape features. Figure 9.6 shows the Born reconstruction and cepstral reconstruction for this image data.

Combination of the wedge shape and high permittivity has made this object very difficult to image. The cepstral method does a reasonable job in recovering the quantitative description of IPS010; however, the shape estimation is still poor both in the Born and cepstral methods. One reason for this poor shape could be the insufficient sampling rate. The same number of illumination angles, which are used to reconstruct simple cylindrical objects, may not be sufficient to reconstruct a wedge-shaped object with sharp features.

As before, the PDFT algorithm was applied to IPS010 to improve reconstruction quality. It is important to have some knowledge about the object, such as size, shape and location, in order to distinguish between target and artifacts. Unwanted signal components can be minimized or eliminated by carefully choosing a prior to encompass V as tightly as one can reasonably estimate without cutting in to the actual dimensions of V. Actually, if one did, the energy of the PDFT estimate would become very large and one can use

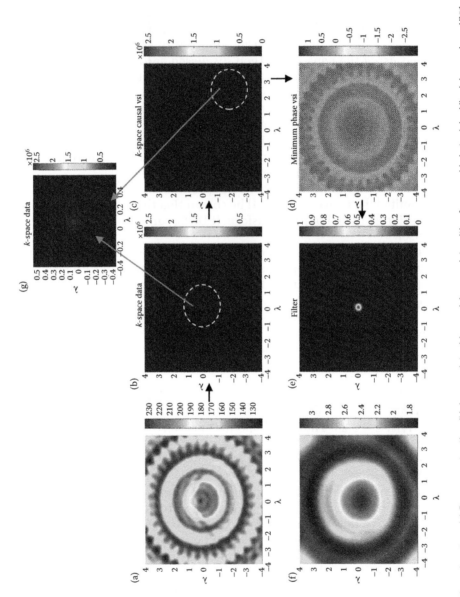

Figure 9.3 Cepstral inversion steps: (a) Born reconstruction, (b) k-space data, (c) causal k-space data with reference added at origin, (d) minimum phase $V(\Psi)$, (e) Gaussian filter, (f) cepstral reconstruction, and (g) zoomed k-space data.

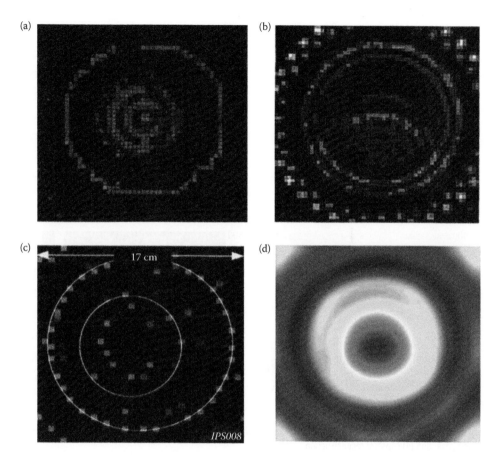

Figure 9.4 IPS008 reconstructions: (a) Reconstruction taken from Byrne and Fitzgerald (1984). (b) Reconstruction taken from Belkebir and Saillard (2001). (c) Reconstruction taken from Estatico et al. (2005). (d) Cepstral reconstruction from Shahid (2009).

this as a means to "shape" or determine the perimeter of an unknown V. For IPS010, various prior functions were investigated. It was determined that the best reconstruction was obtained when a rectangular prior was chosen centered on the center of wedge. Figure 9.7 shows the prior and resulting PDFT estimate of ISP010.

Figure 9.5 IPS008 cepstral reconstruction using PDFT.

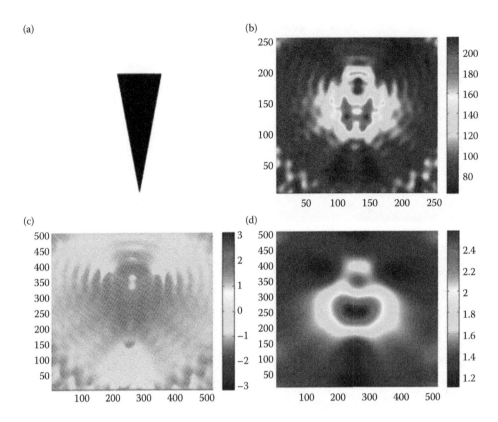

Figure 9.6 IPS010 target configuration: (a) Target object IPS010 geometry, (b) Born reconstruction, (c) minimum phase, and (d) cepstral reconstruction.

9.2 INSTITUT FRESNEL DATA RESULTS

The Institut Fresnel data come from a series of laboratory controlled experiments performed at the Institut Fresnel (Belkebir and Saillard, 2001, 2005), with the same idea of providing a means to help groups evaluate and improve their inverse scattering algorithms. The experimental setup consists of a fixed transmitting antenna and a moving receiving antenna which can move on a circular rail around the object corresponding to the Cartesian coordinate system as shown in Figure 9.8. The transmitting antenna illuminates the target(s) from various locations equidistant around the object. The antennas are located at a distance of 1.67 m from the center of experimental setup. More details about these experimental data can be found in Ayraud et al. (n.d.).

Figure 9.9 shows the schematic representation of the cross sections of the actual target objects with respect to the source and receiver. The scattered field was provided for a range of illumination frequencies and angles. For FoamDielInt and FoamDielExt (the notation for data sets), the emitting antenna was placed at eight different locations which were 45° apart, whereas for FoamTwinDiel and FoamMetExt, which are more complicated objects, the emitting antenna was positioned at 18 locations with 20° angular intervals. The receiving antenna collected complex data at 1° intervals. The scattering experiment was conducted using nine operating frequencies, which range from 2 GHz to 10 GHz. The data is first normalized and then used to compute

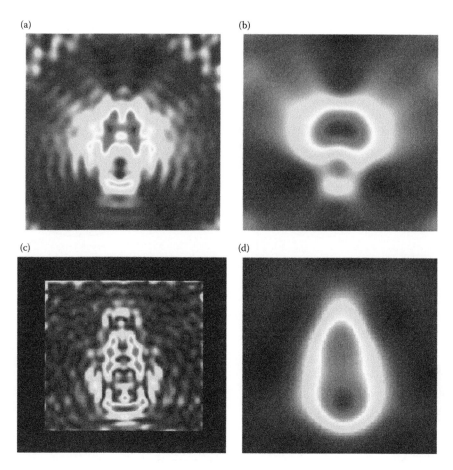

Figure 9.7 IPS010 object: (a) Born reconstruction, (b) cepstral with DFT reconstruction, (c) IPS010 with prior, and (d) cepstral with PDFT reconstruction.

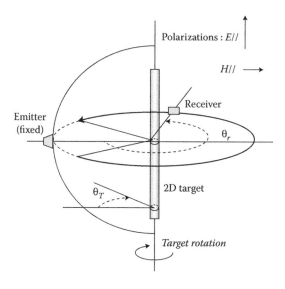

Figure 9.8 Experimental setup from Institut Fresnel.

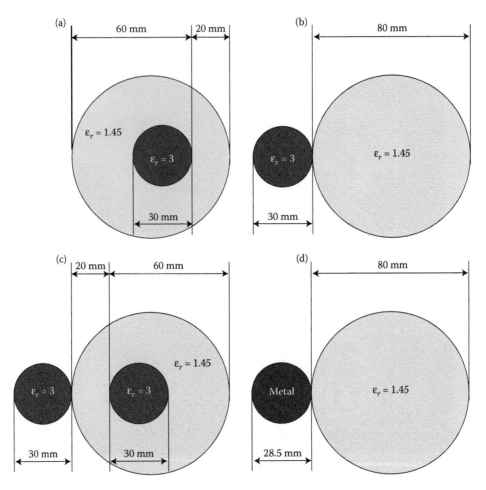

Figure 9.9 Shape and relative permittivity of Institut Fresnel targets: (a) FoamDielInt, (b) FoamDielExt, (c) FoamTwinDiel, and (d) FoamMetExt.

the scattered field by subtracting the complex incident field from the complex total field. The inverse Fourier transform of the scattered field data gives a first Born approximation reconstruction.

9.2.1 FoamDielInt

The FoamDielInt consists of two cylinders: a "foam" of relative permittivity $\varepsilon_r \approx 1.45$ and inside the "foam" there is another circular dielectric of relative permittivity $\varepsilon_r \approx 3.0$. Figure 9.8 shows the reconstruction from the inverse Fourier transform of scattered field data, that is, the first Born reconstruction of FoamDielInt. The Born reconstruction is computed at 6 GHz operating frequency for which the scattering strength of the object is $|kVa| \approx 22$. The object represents a fairly strong scatterer, $|kVa| \gg 1$, as a result of which we see that the first Born reconstruction is not all that good.

Figure 9.10b shows the reconstructed image of FoamDielInt using the Born approximation method. The reconstruction shows several undesirable artifacts and it also fails to give any meaningful (i.e., quantitative) values for relative permittivity. The homomorphic filtering method was then applied to

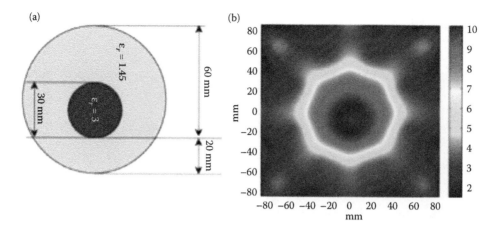

Figure 9.10   FoamDielInt: (a) Actual object, (b) Born reconstruction.

FoamDielInt at 6 GHz. Figure 9.11 shows the implementation of the method to estimate the FoamDielInt object. The available data was made causal by moving it into one quadrant, and then a reference point was added such that the phase of $V\Psi$ lies between $+\pi$ and $-\pi$.

A Gaussian filter was then applied to the cepstral domain, and was then reduced in diameter until no discernable wave-like features associated with Ψ remained in the final image, which is obtained as a further inverse Fourier transform of the filtered cepstrum and exponentiation. Figure 9.11e shows the reconstruction obtained by applying cepstral filtering. The ratio of the contrast of the reconstructed cylinders matches the ratio of the permittivities in the original object. Figure 9.11c shows a log-plot of the cepstrum of $V\Psi$ after the reference point has been added. The log-plot is used to view the low energy features in the cepstral domain. The success of cepstral filtering relies on $V\Psi$ being minimum phase, which is dependent upon the amplitude and location of the "artificial" reference point. Figure 9.12 shows the variation in the quality of reconstruction as we change the reference point amplitude. Figure 9.12c shows that the quality of reconstruction improves as we get closer to satisfying the minimum phase condition. FoamDielInt was also computed for various other frequencies; the results are shown in Figure 9.13.

One of several factors that will affect the quality of the reconstruction (apart from the difficulties associated with inverting multiply scattered data) is the data determined image point-spread function. Figure 9.14 shows the point spread function for different frequencies calculated from the inverse Fourier transformation of the set of delta functions which mark the locations of the sampling points in k-space, as provided by Institut Fresnel. These indicate the extent of features that one can hope to resolve in the final reconstructions of the scattering object $V(r)$ from the measured scattered data in an ideal situation.

At lower frequencies, that is, 3 GHz, the locus of data in k-space is over a much smaller radius Ewald circle than for the 10 GHz case. As a consequence, the main lobe of the point-spread function is much larger at lower frequencies of illumination, and therefore, one can expect to see a lower resolution reconstruction. However, low frequency illumination also implies a better image estimate of $V(r)$ based on the conditions for the validity of the first Born approximation, since a condition for the Born approximation to be valid is

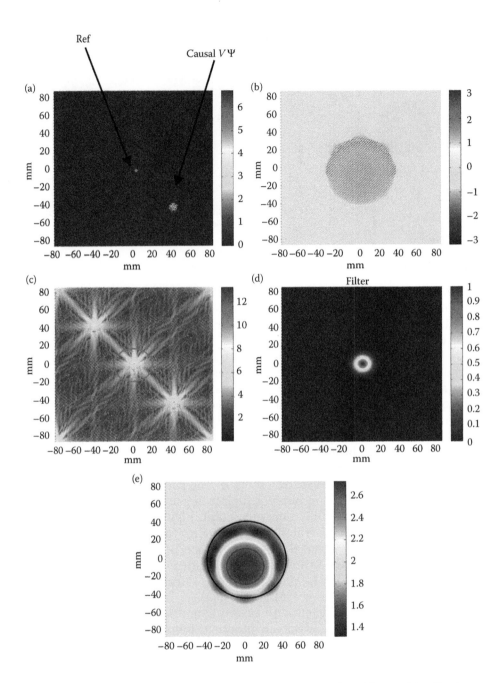

Figure 9.11 Homomorphic filtering applied to FoamDielInt at 6 GHz: (a) Causal k-space data with reference point, (b) minimum phase $V\Psi$, (c) log-plot of cepstral domain, (d) Gaussian filter, and (e) cepstral reconstruction.

that $|kVa| \ll 1$. In addition, Figure 9.14b illustrates that for high frequencies the k-space coverage available from the measured data is insufficient to avoid strong side lobes, which can appear as replicas of the central lobe.

This means that for a reconstruction based on the Born approximation one would expect an optimum performance at some lower frequencies, perhaps including more measured data at a larger number of incident and scattering

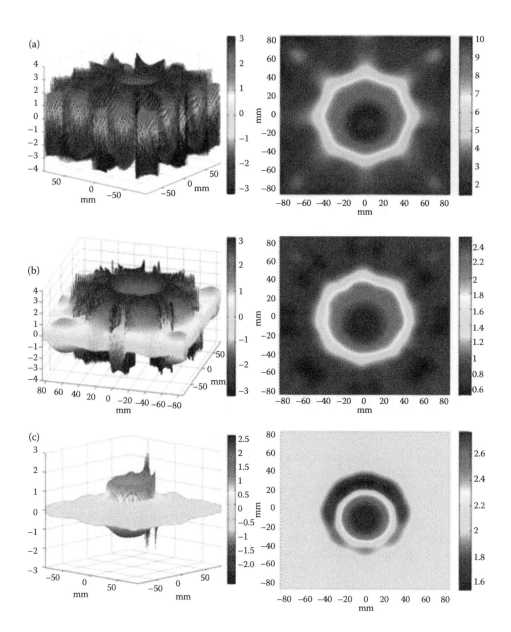

Figure 9.12   Reference strength and cepstral reconstruction. Left side image is the phase of $\nabla\Psi$ and right side image is the cepstral reconstruction: (a) No reference added, (b) a small reference added, and (c) reference amplitude is strong enough such that the phase is less than 2π.

angles. For data collected at 6 GHz, a good trade-off between high resolution and minimum artifacts can be observed.

It is important to note the poor reconstruction and an apparent contrast reversal at 9 and 10 GHz (see Figures 9.13c and 9.13d). It is observed that the contrast of the recovered permittivity of the inner and outer cylinder is reversed, that is, the recovered permittivity of the inner cylinder should be around 3 but it is around 1.7 whereas the outer cylinder should have a permittivity of 1.45 but we recover it around 3. There are two possible explanations

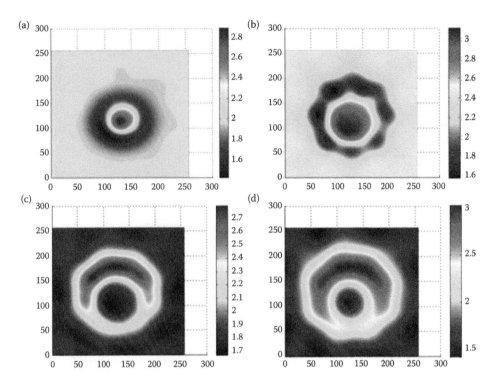

Figure 9.13 Cepstral reconstruction of FoamDielInt at various frequencies: (a) 3 GHz, (b) 7 GHz, (c) 9 GHz, and (d) 10 GHz.

for this. The first is that inversion at lower frequencies meets conditions that increasingly favor the validity of the first Born approximation (i.e. $|kVa| \ll 1$) and for this scatterer at relatively lower frequencies, the sampling rate of the scattered field is adequate. One might also speculate that at higher frequencies, as the effective wavelength is reduced, there may be relatively more scattering from the boundary and less penetration of the field into the target, and so, less scattering from the higher permittivity internal features. Another important factor is that the angular sampling of the scattered field and the number of incident field directions are identical for all frequencies employed. One

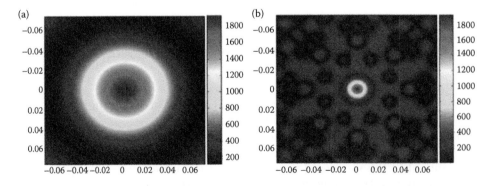

Figure 9.14 Point spread functions corresponding to the locus of k-space scattered field coverage for the frequencies of (a) 3 GHz and (b) 10 GHz.

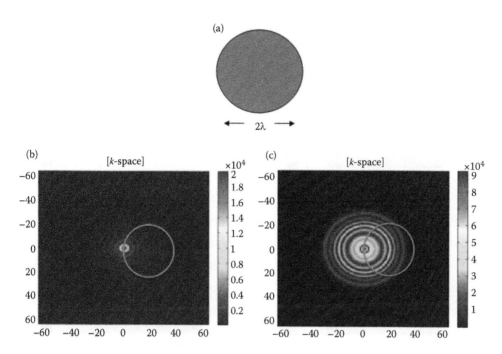

Figure 9.15 The k-space coverage for cylinder of radius 2λ: (a) Target cylinder, (b) k-space coverage for weak scattering cylinder, and (c) k-space coverage for strong scattering cylinder.

can view this as an effective sampling of k-space that results in increasingly spread out high spatial frequency data locations as k increases, as indicated in Figure 9.15, and one might justifiably expect that an increased sampling rate would be necessary in order to capture this information.

Note that for the same geometrically sized cylindrical object in the images above, the k-space coverage varies dramatically as a function of the permittivity of the cylinder. For this reason, small radius Ewald circles will cover mostly low spatial frequencies while large radius Ewald circles will capture information about both low and many high spatial frequencies. The radius of the Ewald sphere in k-space changes with varying incident frequencies. Hence, the mapping between the scattered field data on these circles and the k-space representation of the two different $V\Psi$ shown here results in different regions of k-space being sampled. Small incident k values only map on to lower spatial frequencies. Each estimate obtained from a given incident frequency can add information about different spatial frequencies and, in principle, should help in improving reconstruction quality.

It is also noted that the wrong value for the background permittivity is observed in Figure 9.11e. It has been postulated (Shahid, 2009) that the contrast reversal and incorrect surrounding permittivity level could be caused by the limited sampling of the scattered field for these objects, becoming increasingly limited as k increases for a given source–receiver set of locations. The problem of limited data and undersampling can be addressed by using the PDFT algorithm. Figure 9.16 shows the comparison of DFT and different prior functions incorporated into the PDFT reconstructions. It is evident from Figure 9.16b that the contrast of the reconstruction is incorrect when using

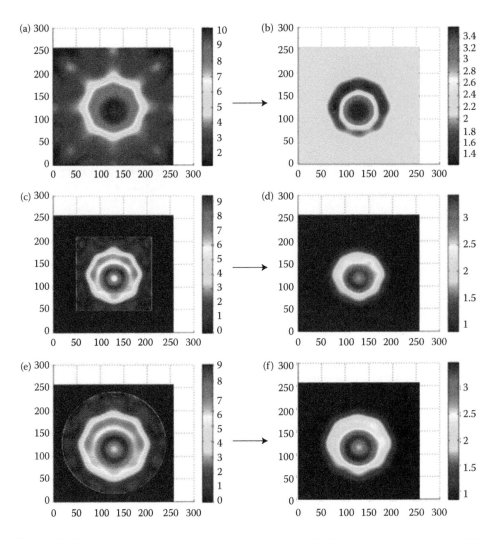

Figure 9.16   FoamDielInt at 6 GHz: (a) Born reconstruction with DFT, (b) cepstral reconstruction using DFT, (c) Born reconstruction using PDFT – a square prior, (d) cepstral reconstruction using PDFT – square prior, (e) Born reconstruction using PDFT – circular prior, and (f) cepstral reconstruction using PDFT – square prior.

cepstral filtering with the DFT, since the background permittivity level has to be the lowest level for this object. Figures 9.16d and 9.16f show that the PDFT has addressed this problem of an incorrect background permittivity. The effect of two different prior functions used with the PDFT is also shown: one a square and the other a circle. The improvement using these is significant, and the choice of prior is not that critical, as expected when it is large compared with V. Considering the fact that FoamDielInt is a strong scattering object, even first Born reconstructions have shown some improvement with the use of PDFT, as is evident from Figures 9.16a and 9.16c.

Since each illumination frequency generates a different $V(\mathbf{r})\langle\Psi\rangle$ with V being the same and $\langle\Psi\rangle$ different for each wavelength, it follows that each illumination frequency provides us with different information and useful redundancy. In other words, changing the frequency of the incident field will

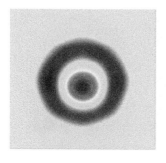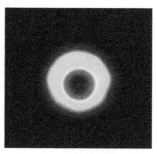

Figure 9.17   FoamDielInt at 6 GHz: (a) Summation of cepstral reconstruction of 4, 5 and 6 GHz with PDFT and (b) summation of cepstral reconstruction of 4, 5 and 6 GHz with DFT.

provide an additional mechanism for extracting spatial frequencies of V alone from the cepstrum. A linear combination of estimates for V acquired in this way will further improve the signal-to-noise ratio (SNR) of the estimate for V while suppressing any residual components from the Fourier transform in each of these images. An example is shown below in Figure 9.17. It is also evident, as expected, that a Gaussian cepstral filter, being apodized, is preferable to a circular or square hard-cut filter.

The summation of cepstral reconstructions for different frequencies when used with the PDFT has shown significant improvement over the summation of cepstral reconstructions with DFT. To compare Figures 9.17a and 9.16f, one sees that the summation of cepstral reconstructions for different frequencies has shown improvement in shape estimation over a single frequency cepstral reconstruction using the P, however summed up cepstral reconstructions still need improvement to recover the permittivity distribution of the object, V. One could feasibly employ a two-step process in which one would use summation for shape retrieval and single frequency reconstruction for quantitative recovery for a more balanced approach.

9.2.2 FoamDielExt

The FoamDielExt consists of a cylinder with of relative permittivity $\varepsilon_r \approx 1.45$ and an external cylinder with relative permittivity $\varepsilon_r \approx 3.0$. The Born reconstruction and cepstral reconstruction at illumination frequency of 6 GHz are shown in Figure 9.18.

It is again observed that the background permittivity level being inconsistent from the actual object's background. The PDFT implementation of FoamDielExt with rectangular prior is shown in Figure 9.19.

In Figure 9.19b the background permittivity from the PDFT estimate is improved significantly as compared to the DFT in Figure 9.18b by the use of a rectangular prior function. The extrapolated values contribute to the improvement of the resolution of the object estimate. Including more prior information in PDFT, such as object feature information or relative permittivity differences, should lead to further improvements in the reconstruction.

9.2.3 FoamTwinExt

FoamTwinExt is one of the most complex objects made available by Institut Fresnel. It is the combination of FoamDielInt and FoamDielExt objects that

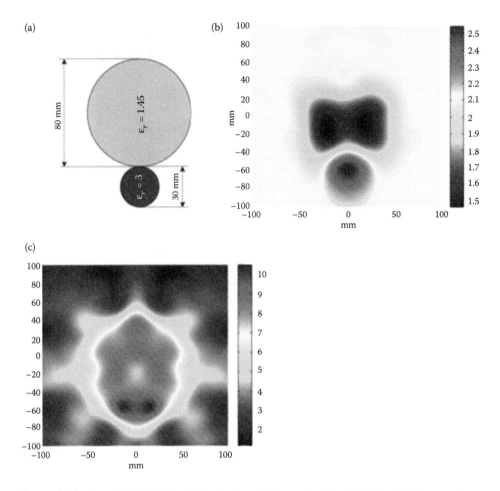

Figure 9.18 FoamDielExt at 6 GHz: (a) Target, (b) cepstral reconstruction with DFT, and (c) Born reconstruction with DFT.

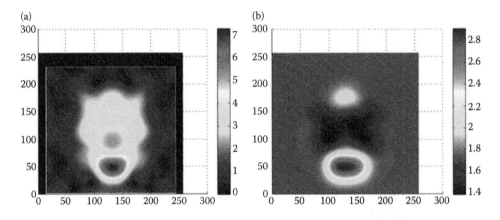

Figure 9.19 FoamDielExt at 6 GHz: (a) Born PDFT reconstruction with rectangular prior and (b) cepstral PDFT reconstruction with rectangular prior.

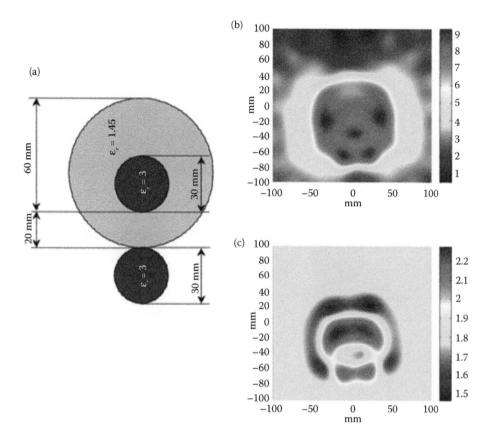

Figure 9.20   FoamTwinExt at 6 GHz: (a) Target, (b) Born reconstruction, and (c) cepstral reconstruction.

is, it contains a large cylinder with relative permittivity of $\varepsilon_r \approx 1.45$, a smaller inner cylinder with relative permittivity of $\varepsilon_r \approx 3.0$ and an outer small cylinder with relative permittivity of $\varepsilon_r \approx 3.0$. The Born reconstruction and cepstral reconstruction for FoamTwinExt are shown in Figure 9.20.

The FoamTwinExt is a complex strong scatterer and a Born reconstruction is not able to retrieve either correct dimensions or the permittivity distribution of object (see Figure 9.20b). Considering the complexity of the object, the cepstral reconstruction has made a reasonable attempt in identifying all three cylinders or permittivity contrast levels as evident in Figure 9.20c. It is possible that the degradation in image quality is due to the lack of measured data, given the extent of multiple scattering that results from the high scattering strength of the original object. The amount of data provided for FoamTwinExt is not sufficient to recover a good image estimate using these techniques. The scattering data for FoamDielInt, which has a simple circular geometry, was collected using 8 incident illuminations, whereas the FoamTwinExt, which has a complex geometry, was collected using only 18 incident illuminations. This is strong evidence to suggest that, as the geometrical and optical complexity of the scattering object increase, one needs higher sampling density of the scattered field or increased degrees of freedom to obtain a meaningful reconstruction of the object.

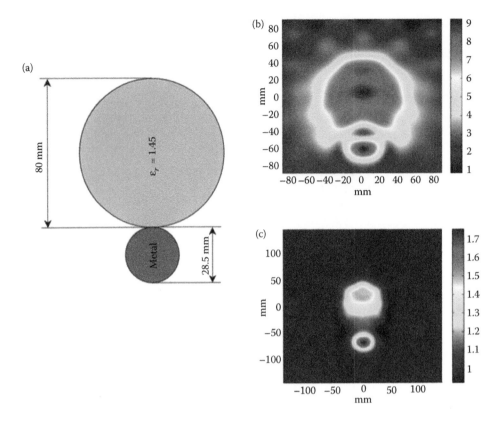

Figure 9.21   FoamMetExt at 6 GHz: (a) Target, (b) Born reconstruction, and (c) cepstral reconstruction.

9.2.4 FoamMetExt

FoamMetExt is a hybrid target which contains a large dielectric cylinder with relative permittivity of $\varepsilon_r \approx 1.45$ and an external metallic cylinder. For metal objects, the incident scattered field is scattered at the surface and then reflects back without entering into the object itself. However, due to the scattered field components, which cross a metal–dielectric interface, the multiple scattering effects become significant. Figure 9.21 shows the reconstruction attempt of the hybrid object using recently discussed techniques. It is evident from the figure that the shape of metal objects can be recovered with good quality even using linear approximations. The Born reconstruction has retrieved reasonably developed shapes for both the outer metal cylinder and the inner dielectric cylinder. Figure 9.21c shows that the cepstral reconstruction has further cleaned the artifacts from the Born reconstruction, but at the expense of losing valid information about the actual dimension of the shapes.

9.3 COMPARISON OF RECONSTRUCTION METHODS

In this section, a comparison of reconstructions from various other methods, some discussed in Chapter 7, that were published in the journal *Inverse Problems* volume 21, issue 6, 2005 is presented, compared and discussed here. These results will also be compared to some of the results from methods presented in this text.

Figure 9.22 shows reconstruction attempts of FoamDielInt using various methods. Figure 9.22a shows the reconstructed image of object using a two-step inexact Newton method (Estatico et al., 2005), where Figure 9.22a is reconstructed at 2 GHz and Figure 9.22b is reconstructed at 5 GHz. Both the reconstructions show artifacts, and the quality of reconstruction seems poor. The recovered permittivity values are far off from the actual values. Figure 9.22c shows reconstruction from an iterative regularized contrast source inversion (CSI) method (van den Berg et al., 1999). The reconstruction quality is average, and it fails to give a good quantitative estimate of relative permittivity.

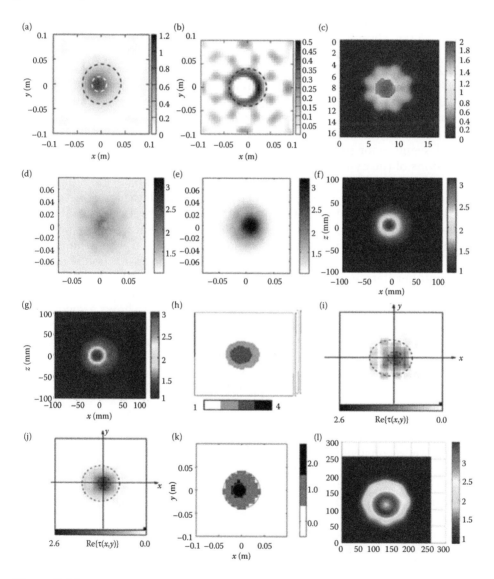

Figure 9.22 Comparison of reconstruction algorithms on FoamDielInt: (a) The two-step inexact Newton method (2 GHz), (b) the two-step inexact Newton method (5 GHz), (c) CSI Method, (d) MGM method, (e) MGM method with adaptive multiscale, (f) DTA/CSI methods, (g) DTA/CSI methods, (h) Bayesian inversion method, (i) IMSA (plane wave) method, (j) IMSA (line source) method, (k) general iterative method, and (l) cepstral method.

Figures 9.22d and 9.22e show the reconstructions obtained using a modified gradient method (MGM) for the inversion of the scattered field data in conjunction with an adaptive multiscale approach based on spline pyramids to improve image quality (Baussard, 2005). Figure 9.22d shows the reconstruction using MGM by itself. Again the quality of image reconstruction seems poor, as the boundaries of inner and outer cylinders are not distinguishable. Figure 9.22e shows reconstruction using MGM with an adaptive multiscale approach (Baussard, 2005). This method seems to do a decent job in recovering permittivities of the cylinders. The reported permittivities for the outer cylinder and the inner cylinder are $\varepsilon_r \approx 2.5$ and $\varepsilon_r \approx 1.68$, respectively. The reconstruction obtained with this approach has shown good results but at the cost of 15% more computation time as compared to MGM. Figure 9.22f,g shows the reconstruction of FoamDielInt by a technique, which combined diagonal tensor approximation (DTA), and CSI (Abubakar et al., 2005). The reconstructions from this method seem by far the best, but again it uses an iterative approach, which is computationally expensive, and there is no guarantee of the convergence of the algorithm to a solution, right or wrong. Figure 9.22 h shows reconstruction from another iterative method based on a Bayesian inversion method. The quality of this reconstruction method seems poor in a sense that it not only fails to retrieve the correct dimensions of the cylinders but also gives a poor estimate of relative permittivity. The reconstructions shown in Figures 9.22i and 9.22j are obtained by using an iterative multiscaling approach (IMSA), which exploits the scattered field data through a multistep reconstruction procedure (Donelli et al., 2005). Figure 9.22i shows the reconstruction when the incident wave is modeled as a plane wave and Figure 9.22j shows the reconstruction when the incident wave is modeled as a line source. Figure 9.22 k shows an image estimate using another iterative approach (Litman, 2005). This approach seems to do a good job in recovering the shape of the object but it fails to recover any quantitative information about the object. Figure 9.22l shows a reconstruction using the cepstral filtering method (Shahid, 2009). Considering the fact that it is a noniterative low computational cost algorithm, the reconstruction not only gives a good estimate of the object's geometry but also gives a meaningful recovery of relative permittivities. The reconstruction comparison for FoamDielExt, FoamTwinDiel, and FoamMetExt are shown in Figures 9.23 through 9.25, respectively.

Figure 9.23 shows the reconstructions of FoamDielExt from various methods as previously. Again most of the methods seem to fail to do a reasonable job in reconstructing FoamDielExt except reconstructions shown in Figures 9.23f and 9.23 g, which was done using a combination of DTI and CSI (Abubakar et al., 2005). Figure 9.23h, which was reconstructed using a Bayesian inversion method (Feron et al., 2005), seems to give a good estimate of shape but seems to lacks quantitative accuracy. Figures 9.23i and 9.23j, which are based on IMSA (Donelli et al. 2005), show artifacts in reconstruction. The reconstruction shown in Figure 9.23e is based on MGM along with a multiscale approach (Baussard, 2005). The quality of reconstruction seems good in a sense that it has not only recovered shape but also relative permittivity. The only downside is that it is an iterative process and it takes 15% more iterations as compared to MGM. The reconstruction from the cepstral method, Figure 9.23l, seems to have done a reasonable job in recovering permittivity

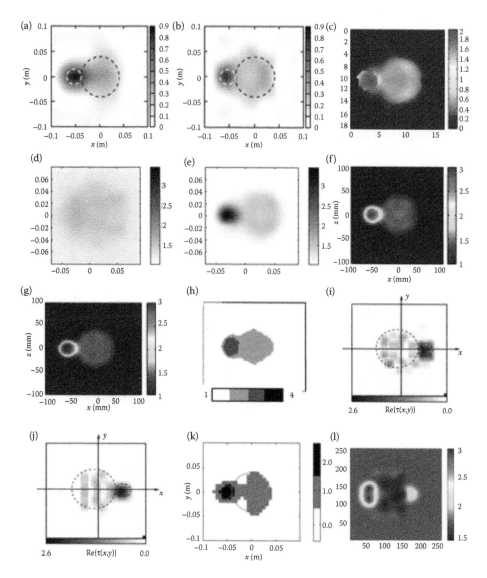

Figure 9.23   Comparison of reconstruction algorithms on FoamDielExt (a) the two-step inexact Newton method (2 GHz), (b) the two-step inexact Newton method (5 GHz), (c) CSI Method, (d) MGM method, (e) MGM method with adaptive multiscale, (f) DTA/CSI methods, (g) DTA/CSI methods, (h) Bayesian Inversion method, (i) IMSA (plane wave) method, (j) IMSA (line source) method, (k) general iterative method, and (l) cepstral method.

and shape of object, but there still seems to be image artifacts present associated with limited data availability.

Figure 9.24 shows the comparison of the reconstructions between various published methods for FoamTwinDiel as before. Because of its complex geometrical configuration, this object is one of the most challenging objects provided by Institut Fresnel to image. Other than the DTI/CSI Method (Abubakar et al., 2005), most of the reconstructions seem to show a low quality estimate of the object dimensions and permitivity. Being noniterative in nature, the cepstral method depends on data coverage. For this multilayered object, in order to get a meaningful reconstruction from the cepstral method it appears

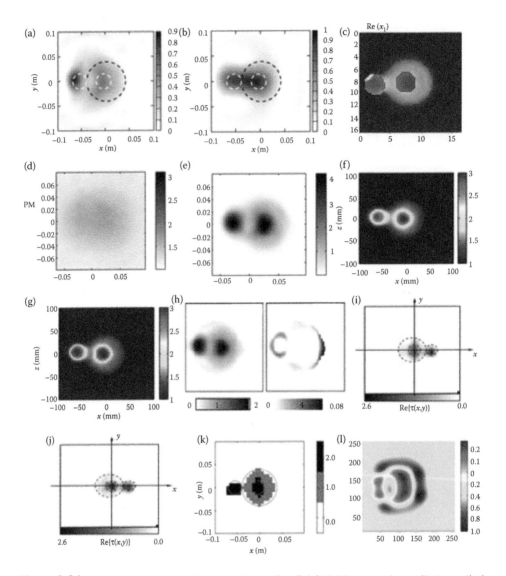

Figure 9.24 Comparison of reconstruction algorithms on FoamTwinDiel: (a) two-step inexact Newton method (2 GHz), (b) two-step inexact Newton method (5 GHz), (c) CSI method, (d) MGM Method, (e) MGM method with adaptive multiscale, (f) DTA/CSI methods, (g) DTA/CSI methods, (h) Bayesian inversion method, (i) IMSA (plane wave) method, (j) IMSA (line source) method, (k) General Iterative method, and (l) cepstral method.

that more data is needed. Even with the limited amount of data, the cepstral method still seems to be able to isolate contrast difference between all three cylinders.

Figure 9.25 shows the comparison of different reconstruction methods for FoamMetExt data. The two-step Newton method (Estatico et al., 2005), which attempted to reconstruct the first three objects, was not able to show any reconstruction for the metal object. Also the iterative method proposed in Litman (2005) did not present any reconstruction for the metal object. Figure 9.25a based on the CSI Method (van den Berg et al., 1999) seems to show good reconstruction for the shape, but does not seem to provide any quantitative description. Figure 9.25d seems to show decent performance for both shape

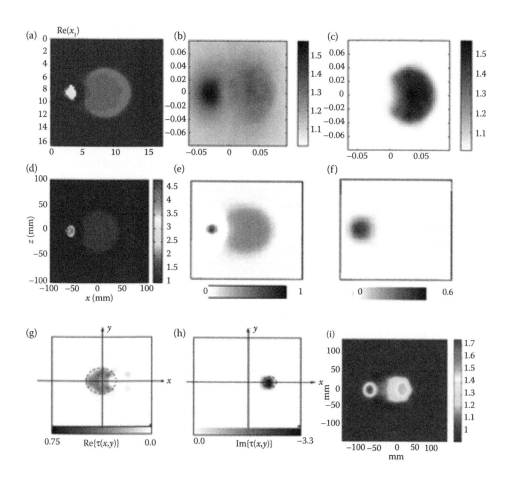

Figure 9.25 Comparison of reconstruction algorithms on FoamMetExt: (a) CSI method, (b) MGM method, (c) MGM method with adaptive multiscale, (d) DTA/CSI methods, (e) Bayesian inversion method (real), (f) Bayesian inversion method (imaginary), (g) IMSA (Real) method, (h) IMSA (imaginary) method, and (i) cepstral method.

and a quantitative image. Figures 9.25e and 9.25f show the real and imaginary part of the reconstructions using the Bayesian inversion method (Feron et al., 2005). Figures 9.25 g and 9.25 h show real and imaginary part of metal object using IMSA (Donelli et al., 2005). Both the reconstructions from Feron et al. (2005) and Donelli et al. (2005) seem to lack quantitative recovery. In addition, they seem to fail to recover all object features in a single image. The image from the cepstral method seems to show good shape and permittivity estimation, although it still seems to suffer from incorrect background permittivity, which is likely due to limited data availability.

9.4 FINAL REMARKS AND SUMMARY

One might ask why the reconstructions in the previous using various methods applied to measured data are relatively poor. This is to some extent the case even for weakly scattering objects. If we look more carefully at the data provided by AFRL or the Institut Fresnel, these were made available to assist with the development and improvement of inverse scattering algorithms, but, before guidelines were known, of the kind we have provided here, relating to degrees of freedom.

In Chapter 6.1 it was stated in Equation 6.1 that

$$N_{\text{3-D}} = \frac{B_V \cdot n_{\max}}{\lambda^3}$$

where

$N_{\text{3-D}}$ is the minimum degrees of freedom required in 3 dimensions
B_v is the target volume
$n_{\max}$ is the maximum index of refraction
λ the wavelength

which could be modified as in Equation 6.2 for 2-D as

$$N_{\text{2-D}} = \frac{A_V \cdot n_{\max}}{\lambda^2}$$

where the target volume is replaced with the target area A_v.

From the many simulations presented here, we can conclude how the number of degrees of freedom provides a metric for the number of data points necessary to recover a reasonable image. In the figures shown in Chapter 6, the number of degrees of freedom, N, is given and then the images are shown as function of $\varepsilon(r)$ and the number of data used, expressed both as a function of N and as an absolute number given the objects' dimensions and permittivities. We noted that there are obviously many factors that are affecting the quality of these reconstructions, not the least being the fact that even using a cepstral filter, though the filter shape has not been optimized. Nevertheless, a reasonable sense of how many measurements must be made can be estimated and at least $4N^2$ appears to be necessary. One can consider this to represent the fact that there needs to be at least $2N$ source locations and at least $2N$ receiver locations. However, in practice, we have found that assigning more degrees of freedom to the receiver locations rather than the source locations appears to be slightly more beneficial. We cannot rule out that a statement such as this might be object-dependent.

Take the IPS008 target which is discussed in Section 9.1.1 above, which was considered a strongly scattering object since $k|V|a \approx 87$ where k, the wave number, is calculated as $k = 2\pi/\lambda = 2\pi/0.03 = 209.5 \text{ m}^{-1}$, V is the scattering strength or average permittivity and "a" is the dimension of the largest feature of object. The 2-D number of degrees of freedom can be estimated to be $\pi(5\lambda)^2$ $3^{1/2}/\lambda^2 \sim 45$. The value of $4N^2 \sim 8000$. The measured Ipswich data consisted of 36 illumination directions, at equal angular separations of $10°$ and 180 complex scattered field measurements for each view angle using a frequency of 10 GHz and hence the total number of data available is of the order 6500 which we argue could well be insufficient. This being the case, then even a perfect inverse scattering algorithm would not be able to generate an image that is a complete rendition of the original object.

Let us consider the measured data provided by Institut Fresnel, where the models are defined in Figure 9.9. The FoamDielInt consists of two cylinders: a "foam" of relative-permittivity $\varepsilon_r \approx 1.45$ and inside the "foam" there is another circular dielectric of relative permittivity $\varepsilon_r \approx 3.0$. From Chapter 5, Table 5.1 shows the reconstruction from the inverse Fourier transform of scattered field data, that is, first Born reconstruction of FoamDielInt. The Born reconstruction is computed at 6 GHz operating frequency for which the scattering strength of the object is $|kVa| \approx 22$. The object represents a fairly strong scatterer, $|kVa| \gg 1$, as a result of which we see that the first Born reconstruction is not all that good. For FoamDielInt and FoamDielExt, the emitting antenna was placed at eight different locations which were 45° apart whereas, for FoamTwinDiel and FoamMetExt, which are more complicated objects, the emitting antenna was positioned at 18 locations with 20° angular intervals. The receiving antenna collected complex data at 1° intervals. The scattering experiment was conducted using nine operating frequencies, which range from 2 GHz to 10 GHz.

The number of degrees of freedom for one of these objects is approximately given by 11×11 cm$^2 \times 1.8/\lambda^2$, and the wavelength varies from 15 cm to 3 cm depending on the frequency used. This gives $1 < N < 25$ assuming no metal is included. However, it is the highest frequency case that will dictate whether the number of measurements taken is sufficient. At lower frequencies, the data collection might well be adequate but the resolution limit using 2 GHz illumination is approximately the size of the entire scattering object, revealing only whether it is there or not; not very useful in our opinion. If we consider $N = 25$, then $4N^2 = 2500$. Assuming 18 antenna positions and measurements of the scattered field for each and every degree, the total number of data available are 6300. This should be sufficient for the nonmetallic objects. For the choice of only eight incident field directions, 2800 is just about sufficient. Consequently, we would argue that we can have some confidence in algorithms that recover good reconstructions from the Institut Fresnel data of the sets of nonmetallic objects.

Finally, in Section 6.4, we discussed imaging resonant objects and noted the improved appearance of the reconstruction of cylinders when illuminated at frequencies close to their Mie resonant frequencies. Figure 9.26 shows on the left a reconstruction of a cylinder with large permittivity, the black circle

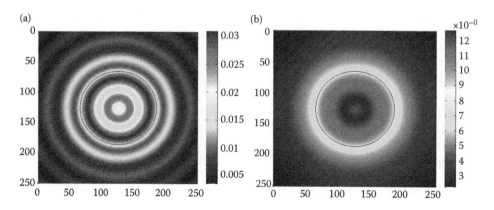

Figure 9.26   (a) An image of a cylinder with large permittivity making the first Born approximation invalid. (b) More uniform reconstruction of the cross section of a cylinder for a weaker scatterer with $\varepsilon_r = 1.1$. The left side image, as expected, has approximately the correct diameter.

defining its actual diameter. On the right we see, to the same scale, an image of a weaker object (permittivity = 1.1) for which the Born approximation is not good, but not necessarily that bad either. The detrimental effect on the quality of the reconstructed image is evident by the degree to which the periodic features arise in the image. This degradation is due to the data truncation in k-space. As the permittivity of the cylinder increases, we pass through a series of resonances and for each of these, there is a strong field enhancement inside the cylinder and the measured scattered field in the far field will reflect this. We see that for those situations, the Born reconstructions are quite good. This is evident by matching the ε_r for maximum scattering in Figures 6.3 and 6.4 with the corresponding reconstructions shown in Tables 6.11 and 6.12, respectively. What is worth noting about these sets of reconstructions is to see how, at resonance, the far-field measurements seemingly correspond to a relatively weak scatterer but with a larger diameter. Consider for example the reconstructions in Table 6.12 for $\varepsilon_r = 1.4$ compared with $\varepsilon_r = 2.1$; both appear reasonably uniform inside the cylinder but neither of them is the correct diameter. Not unsurprisingly, the relative distortion (increase) in size is proportional to the increased radar cross section. The problem of nonscattering and scattering structures and the so-called "bound states" that can store energy in a scattering object at certain frequencies has long been recognized as an inherent ambiguity or lack of uniqueness one must accept in inverse scattering problems. Of course, changing the frequency of illumination can resolve such ambiguities but not entirely since the material properties such as permittivity will be wavelength-dependent.

REFERENCES

Abubakar, A., vand den Berg, P. M., and Habashy, M. 2005. Application of the multiplicative regularized contrast source inversion method on TM & TE polarized experimental Fresnel data. *Inverse Problems, 21*, 5–13.

Baussard, A. 2005. Inversion of multi-frequency experimental data using an adaptive multiscale approach. *Inverse Problems, 21*, 15–31.

Belkebir, K. and Saillard, M. 2001. Special section: Testing inversion algorithms against experimental data. *Inverse Problems, 17*, 1565–1571.

Belkebir, K. and Saillard, M. 2005. Special section on testing inversion algorithms against experimental data: Inhomogeneous targets, *Inverse Problems*, 21, S1–S3.

Byrne, C. L. and Fitzgerald, R. M. 1984. Spectral estimators that extend the maximum entropy and maximum likelihood methods. *SIAM Journal of Applied Mathematics, 44*, 425–442.

Donelli, M., Franceschini, D., Massa, A., Pastorino, M., and Zanetti, A. 2005. Multi-resolution iterative inversion of real inhomogeneous targets. *Inverse Problems, 21*, 51–63.

Estatico, C., Bozza, G., Massa, A., Pastorino, M., and Randazzo, A. 2005. A two-step iterative inexact-Newton method for electromagnetic imaging of dielectric structures from real data. *Inverse Problems, 21*, 81–94.

Feron, O., Duchene, B., and Mohammad-Djafari, A. 2005. Microwave imaging of inhomogeneous objects made of a finite number of dielectric and conductive materials from experimental data. *Inverse Problems, 21*, 95–115.

Litman, A. 2005. Reconstruction by level sets of n-ary scattering obstacles. *Inverse Problems, 21*, 131–152.

Maponi, P., Misici, L., and Zirilli, F. (1997, April). A numerical method to solve the inverse medium problem: An application to the Ipswich data. *IEEE Antennas and Propagation Magazine, 39*(2), 14–19.

McGahan, R. V. and Kleinman, R. E. 1999. The third annual special session on image reconstruction using real data. Parts 1 and 2. *IEEE Antenna Propagation Magazine, 41*(1/2), 34–51/20–40.

Shahid, U. 2009. *Signal Processing Based Method for Solving Inverse Scattering Problems.* PhD Dissertation, Optics. University of North Carolina at Charlotte, Charlotte: UMI/ProQuest LLC.

van den Berg, P. M., van Broekhoven, A. L., and Abubakar, A. 1999. Extended contrast source inversion. *Inversion Problems, 15*, 1325–1344.

Ten

Advanced Cepstral Filtering

10.1 INDEPENDENT PROCESSING OF SOURCE DATA

It has been shown in a previous work that improvements can be made to a reconstructed image produced by using the Born approximation method by performing filtering in the cepstrum domain (Shahid, 2009). Previously in applying this method, the filtering was applied to the $V\langle\Psi\rangle$ image data from the combined sources. In retrospect, while the expectation was that the field term might appear noise-like, a more rigorous approach is described here. The basis of the derivation in Chapter 4, in particular Equations 4.28 through 4.31, is that it is for a single incident wave or direction. Since this is the case, it seems logical that the processing of the data should be carried out on a source-by-source basis and then recombined later in the process. When the independent source data is combined at the beginning, this gives the opportunity for the information relating to the unwanted Ψ terms to be combined and possibly modulated in such a way that parts of its spectrum may be even more difficult to locate and/or remove.

This being the case, the cepstrum filtering method in Shahid (2009) has been re-implemented on a source-by-source basis in lieu of a combined source method. To accomplish this, the Ewald circle of data for each independent source was derived separately and the Born image for each source was computed as illustrated in Table 10.1 for two different target sets, a circle and two triangles. The data for each source was then transformed into the cepstrum domain and filtered accordingly as before in Shahid (2009), again as illustrated in Table 10.2 for these two different targets sets. The question then arises at to what is the optimum step in the reconstruction process to recombine the independent data into a cumulative data set from which the entire image could be recovered. There are two opportunities to do this. The first is in the cepstrum domain directly after filtering. The second is to transform each of the independent filtered source data back into the image domain and recombine there. Both of these scenarios have been implemented and will be examined.

This total process was performed on four unique target sets to provide a broad comparison of the general performance of these methods. Four target sets with various permittivities were modeled and filtered as described above (Table 10.3). As seen in Table 10.3, it is obvious that all of the cepstrum filtered images show improvement over the image obtained using just the Born method, both in regard to image boundaries as well as noise reduction. In the analysis of these methods, it has been observed that as the permittivity of a target is increased, the apparent size of the resulting target reconstruction using the Born approximation method is artificially inflated in relation to the actual target size. This can be seen in the images shown in Table 10.3.

Table 10.1   Sequence of Images Showing the Individual Source Ewald Circles and the Corresponding Born Images for Each Source

Ewald Circles for Each of the 9 Sources

Composite Images

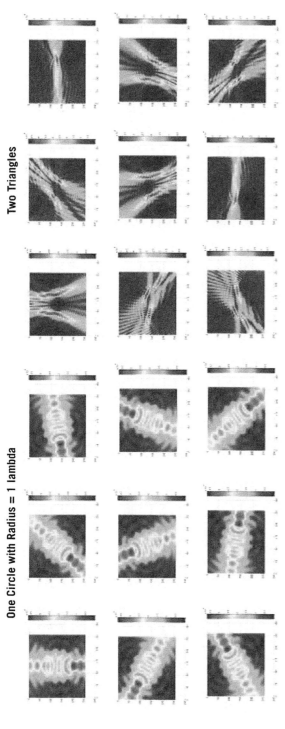

Two Triangles

One Circle with Radius = 1 lambda

Corresponding Born Image for Each Ewald Circle Above

Note: The composite of the Ewald circles and the composite Born reconstructions are also shown for a circle and for two triangles.

Table 10.2 Sequence of Images Showing the Individual Born Images for Each Source and the Corresponding Cepstrum Filtered Image

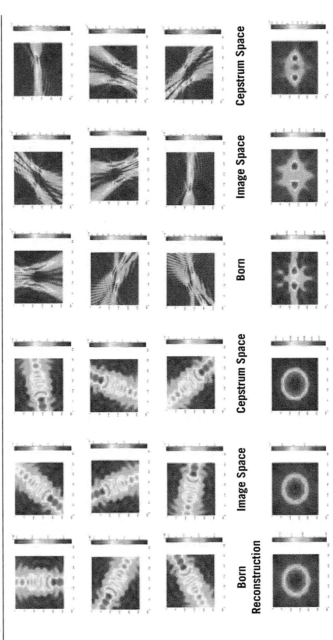

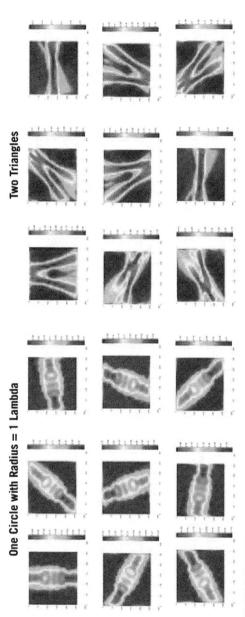

Note: The resulting Cepstrum composite images are also shown for the Born images, sources combined in Cepstrum space and image space.

Table 10.3 A Comparison of Reconstructed Images for Various Targets Using the Born Method, Cepstrum Filtering of the Born Reconstructions, and Processing Sources Independently and Recombining in Image Space, and Cepstrum Space

Born Reconstruction with 12 Sources @ 5 GHz and 120 Receivers	Cepstrum of Born Reconstruction	Processed Independent Sources Combined in Image Space	Processed Sources Combined in Cepstrum Space

1 Circle
Radius = λ
er = 1.8

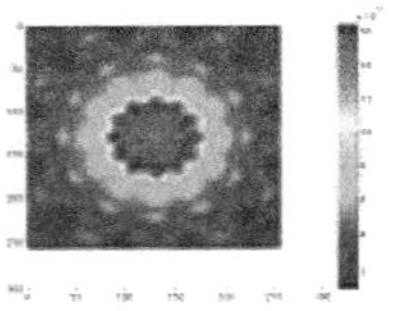

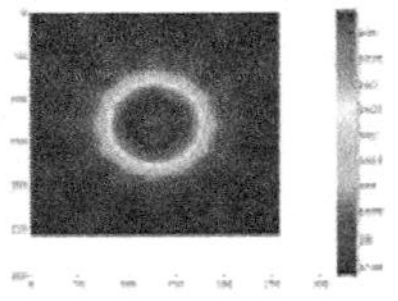

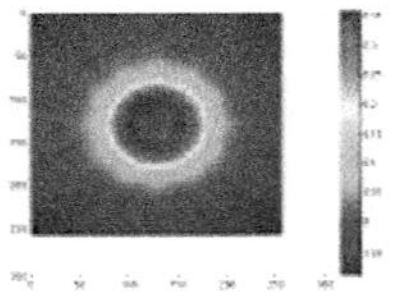

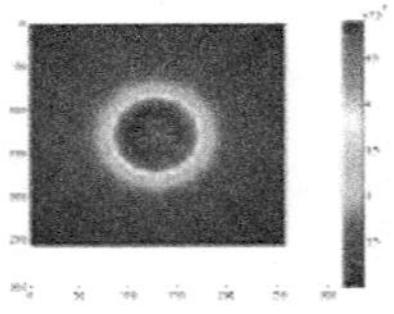

2 Circles
Radius = λ
er = 1.5

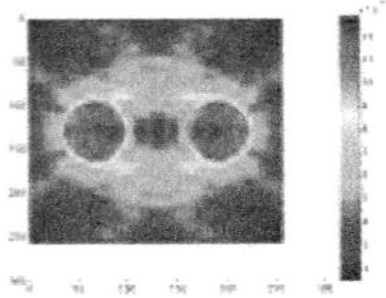

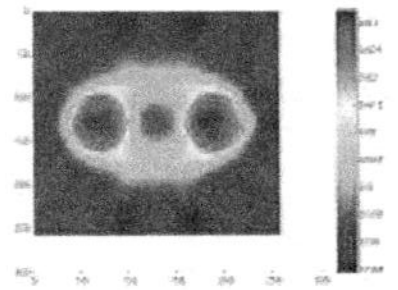

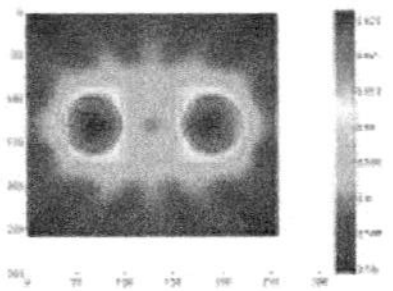

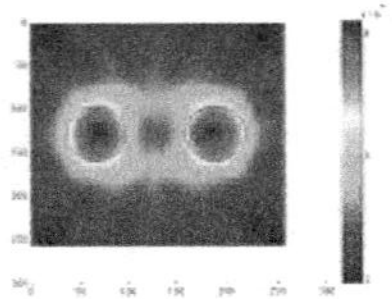

2 Squares
Sides = 2λ
er = 1.4

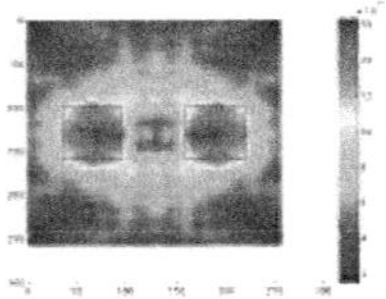

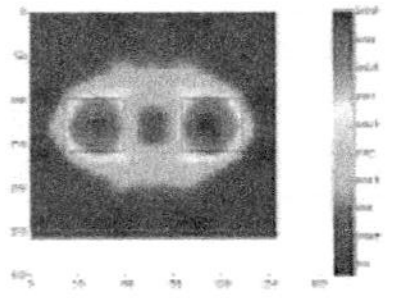

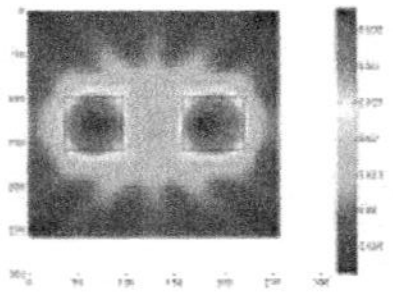

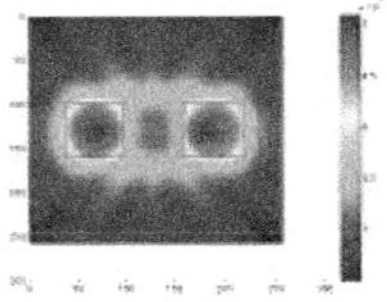

2 Triangles
Height = 3λ
er = 1.5

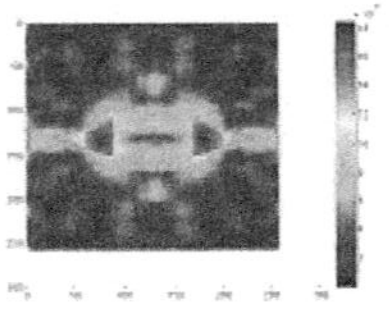

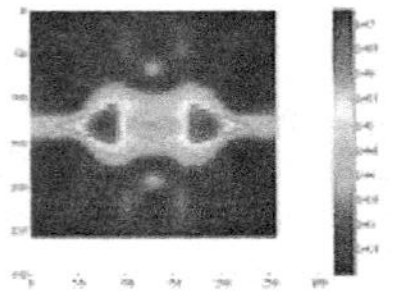

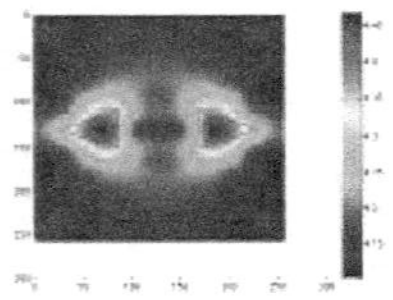

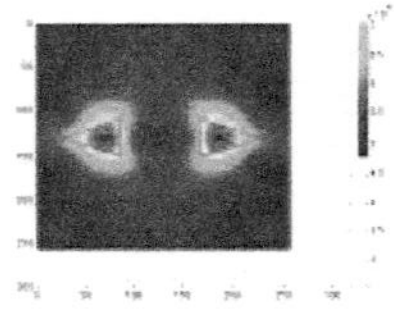

It appears in the filtering processes where the sources are independently filtered, showing that this size inflation is decreased. Moreover, it appears that the images from the process of where the independent source data were combined in the cepstrum domain seem to perform slightly better in this regard. Additionally, the independent source filtered images exhibit a scaling issue as compared to the Born image and the cepstrum-filtered Born images. The image with the data combined in cepstrum space has a magnitude or scale that decreases as the number of sources is increased, while the image with the data combined in image space has a magnitude that increases as the number of sources is increased. It will be shown in a later section that a scaling factor, dependent on the number of sources, can be applied to adjust or improve these inappropriately scaled values. Taking all this into account, and analyzing the images in Table 10.3, it appears that in general, the method of processing the independent source data separately followed by combining these data sets in the cepstrum domain appears to perform better than the other methods examined.

10.2   EFFECTS OF MODIFIED FILTERS IN CEPSTRAL DOMAIN

As discussed earlier, it is known and understood that in the 2-D cepstral filtering method, it is necessary to use some type of low pass filter to eliminate or attenuate the Ψ component in the data; there is very little material available to assist in determining the optimal parameters for this filter. This is compounded even further in that there is little information for one to intuitively know where the "signal" data and "noise" are located in the cepstrum space since the cepstrum domain does not have a linear relationship to the spectral domain. This being the case, some experimentation may be necessary to gain a better understanding of this domain to help in creating the optimum filter for better image reconstruction. It was demonstrated in Shahid (2009) that a Gaussian type filter performs reasonably well in filtering in the cepstrum domain, and this was characterized by the following equation:

$$\text{Filter} = \frac{1}{\sqrt{2\pi}\sigma} e^{-[x^2 + y^2]/2\sigma^2} \qquad (10.1)$$

The σ term in Equation 10.1 dictates how wide the filter will be, which we will refer to as its bandwidth. Conventional filter theory would suggest that it is desirable to have the filter bandwidth to be as wide as possible to allow as much of the desirable signal spectrum to pass as possible, while at the same time narrow enough to significantly attenuate the undesirable signal (or noise) as much as possible. This assumption is based on the noise having mostly higher spatial frequencies than those associated with the original target, $V(\mathbf{r})$, which need not be necessarily the case, especially in the cepstrum domain. Therefore, it seems a logical place to start to vary the width or σ term and observe the effects, if any, on the performance of the various filtering methods described above. Table 10.4 contains a family of images that show how each of the cepstrum filtering methods mentioned in the previous section performs using a Gaussian filter described by Equation 10.1 with gradually increasing width or σ term.

Table 10.4 A Comparison of Reconstructed Images for Triangle Targets Using the Combined Source Cepstrum Filtering of the Born Images, and Processing Sources Independently and Recombining in the Image Space and also the Cepstrum Space for Increasing σ

Filters	36 Combined Source	Ind. Sources in Image Space	Ind. Sources in Cepstrum Space
3			No valid output for reconstructed image
5			No valid output for reconstructed image
10			No valid output for reconstructed image
15			

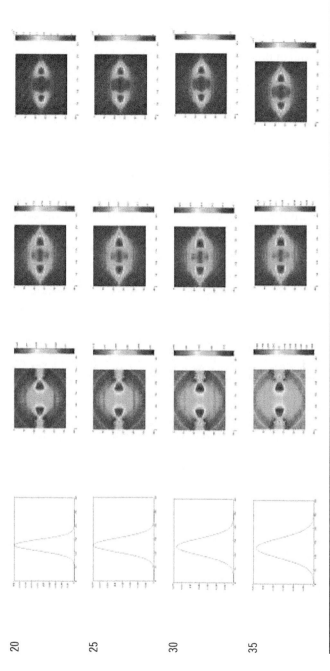

20

25

30

35

Note: The triangles have a height of 3λ and $\varepsilon_r = 1.5$.

It is clear from the images in Table 10.4 that there is no significant gain or benefit to have a width for σ greater than about 10 for the particular family of targets considered here. It should also be noted that there was no valid reconstructed image output for the "combined in cepstrum space" filtering method for values of σ of 3 and 10. It will be shown later that this is due to the peak value of the filter, which can be addressed by scaling. However, it should be noted that this form of Gaussian filter has a maximum value at the center that varies inversely as σ varies to maintain a value of 1 for the total area under the curve. The larger the value of σ is the smaller the peak of the filter. It again will be shown later that this plays a very significant role in the filtering process in the cepstrum domain.

Now that it is known that the majority, if not all, of the "good" or desirable part of the spectrum lies within $\sigma = 10$ of the center of the filter, the next question is whether there is any location within this region that has "bad" or undesirable data. To investigate this, a series of modified Gaussian filters are used to try and further characterize the mapping of data in the cepstrum domain. To do this, a Gaussian filter with a σ value set to 30 was strategically "notched" with gaps of varying widths and locations to see if there was any location "under" the Gaussian filter that is more important than any other for the "good" or the "bad" data.

The results of this filter experiment and its effects on the various types of cepstrum filtering discussed previously are shown in Table 10.5. It is evident again from this table and from Table 10.4 that the important signal data is located within the $\sigma = 10$ range. There does not seem to be any benefit, in general, in trying to include any data from the outer limits of the cepstrum domain. It is demonstrated clearly from the last set of images in Table 10.5, where the $\sigma = 0$ to $\sigma = 5$ range is filtered out that this is where the most important information is since the reconstructions for this condition are very poor, and the original target image is undetermined even with 36/360 degrees of freedom for sources/receivers, respectively.

Now that the optimal value for σ has been investigated and evaluated for the Gaussian filter in regard to the bandwidth, we will now investigate the peak value. It is observed in Table 10.4 that the "combined cepstrum" method does not produce any results for three of the scenarios presented. It was assumed that this was due to the fact that the pertinent "harmonics" were either outside the pass band, or they were excessively attenuated due to the reduced peak that is inherent to the Gaussian function. By using some simple experimentation on the peak value of the Gaussian by using simple scaling techniques, it was discovered that the issue was not that the Gaussian filter peak was too low, but that it was, in fact, too high. The fact is that there is an apparent inverse relationship between the peak value of the Gaussian filter and the overall range in magnitude of the resulting image using the "combined cepstrum" method. Moreover, it was observed through simulations that the minimum constraint for the peak of the filter to insure that the "combined cepstrum" method produces a reconstructed image is that the peak must be equal to or less than $1/N_s$, where N_s is the total number of sources. It was also observed that, in general, to scale the magnitude of the reconstructed image produced by the "combined cepstrum" method in order to be in the same general range as the magnitude of the image produced using the Born reconstruction, it was necessary for the filter to be scaled by an additional value

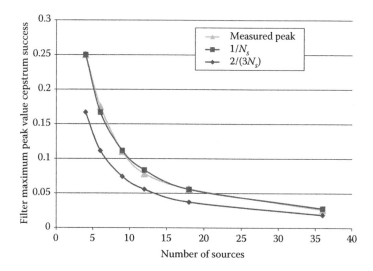

Figure 10.1 Filter peak values versus the number of sources used required causing the "combined cepstrum" method to produce a reconstructed image.

of approximately 2/3. The peak values for each of the sources are shown in Figure 10.1. In this figure, the trend line for this process is drawn and compared with the ideal trend line of $1/N_s$, which demonstrates the conclusion of $1/N_s$ for the minimum constraint for the Gaussian filter for the "combined cepstrum" method. In addition, the line for the "2/3" adjusted line for magnitude adjustment is also shown.

Using the "optimized" value for σ and the new peak scaling factor of $2/(3N_s)$, the process was run again for the two triangles target set for various number of sources using each of the processing methods for comparison of performance. The results of this experiment are shown in Table 10.6. The "optimized combined cepstrum" method appears to consistently perform the best out of all the methods considered.

The final aspect of investigating filtering or removal of unwanted parts of the spectrum or data consists of trying to remove or subtract a cepstrum representation of the incident field, Ψ_{inc}. The reasoning behind this, as discussed in Section 8.1, is that in the cepstrum domain, after the reference has been added, we have a representation of $V\Psi$ in the form of the following approximation:

$$\approx \log\left(\frac{V}{R}\right) + \frac{\Psi}{R'} \tag{10.2}$$

As discussed in Chapter 8, it appears that by subtracting a weighted cepstrum representation of Ψ_{inc} we would then have

$$\approx \log\left(\frac{V}{R}\right) + \frac{\Psi}{R'} - \kappa\Psi_{\mathrm{inc}} \tag{10.3}$$

Table 10.5 A Comparison of Reconstructed Images for Triangle Targets Using the Combined Source Cepstrum Filtering of the Born Images, and Processing Sources Independently and Recombining in the Image Space and also the Cepstrum Space for Various Filters

Gap	Filters	Combined Source	Ind. Sources in Image Space	Ind. Sources in Cepstrum Space
3 to 100				
4 to 100				
5 to 100				
10 to 100				

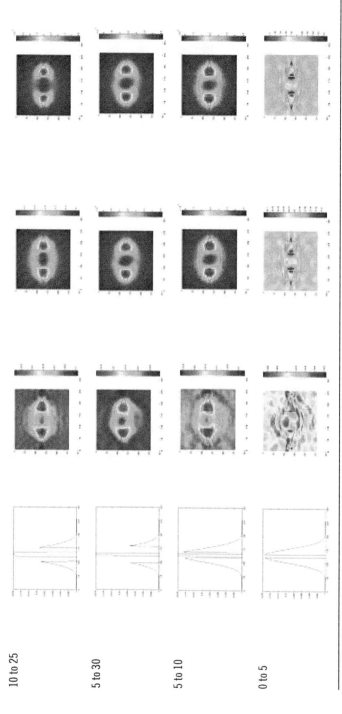

10 to 25

5 to 30

5 to 10

0 to 5

Note: The triangles have a height of 3λ and $\varepsilon_r = 1.5$.

Table 10.6   A Comparison of Reconstructed Images for Triangle Targets Using the Combined Source Cepstrum Filtering of the Born Images, and Processing Sources Independently and Recombining in Images Space, and Cepstrum Space for "Optimized" Filters

N_s	Filter	Born	Cepstrum of Born	Image Space	Cepstrum Space
4					
6					
9					
12					
18					
36					

Note: Images were reconstructed using 36 sources (5 GHz) and 360 receivers all equally spaced. The triangles have a height of 3λ and $\varepsilon_r = 1.5$.

If the factor "κ" can be found that is close to $1/R'$ and $\langle \Psi_{inc} \rangle$ is a close approximation to $\langle \Psi \rangle$, then improvement in the resulting reconstructed image can be expected.

This is what was implemented in the MATLAB® code for both the combined source method and the independent source processing method combined in the cepstrum domain. The effect on the combined source method was observed first. It appears that with a weighting factor of about 4, the reconstructed image does seem to fill more of the original target boundary. Examples of this are shown in Tables 10.7 and 10.8. The independent source method was observed

Table 10.7 A Comparison of Reconstructed Images for Triangle Targets with Dimensions of 3λ Using the Combined Source Cepstrum Filtering of the Born Images for Three Different Permittivity Values

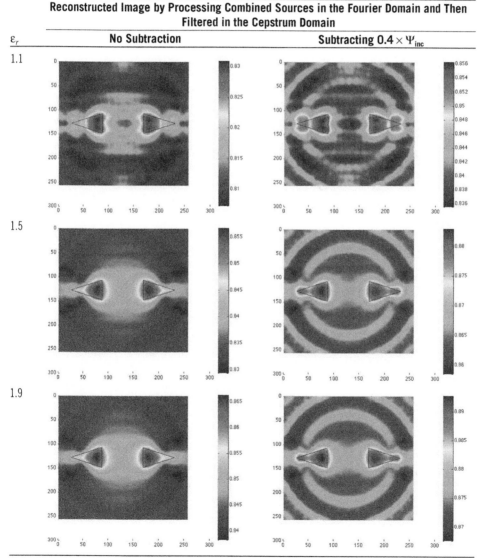

	Reconstructed Image by Processing Combined Sources in the Fourier Domain and Then Filtered in the Cepstrum Domain	
ε_r	No Subtraction	Subtracting $0.4 \times \Psi_{inc}$

Note: These images compare the results of subtracting the cepstrum version of the incident field times a weighting factor of 0.4 ($-0.4 \times \Psi_{inc}$).

Table 10.8 A Comparison of Reconstructed Images for Square Targets with Sides of 2λ Using the Combined Source Cepstrum Filtering of the Born Images for Three Different Permittivity Values

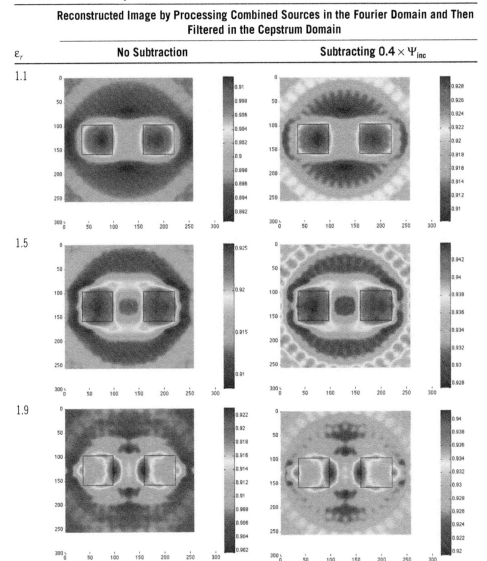

	Reconstructed Image by Processing Combined Sources in the Fourier Domain and Then Filtered in the Cepstrum Domain	
ε_r	**No Subtraction**	**Subtracting $0.4 \times \Psi_{inc}$**

Note: These images compare the results of subtracting the cepstrum version of the incident field times a weighting factor of 0.4 ($-0.4 \times \Psi_{inc}$).

next, and while there was minimal change in the image boundary itself, for a weighting factor of 1, a significant improvement in the scale of the reconstructed image was noted, as shown in Tables 10.9 through 10.12.

10.3 EFFECTS OF RANDOM UNDERSAMPLING

As mentioned in Chapter 6, aliasing can become a significant issue as the number of receivers is reduced and the spacing between the receivers is increased.

Table 10.9   A Comparison of Reconstructed Images for a Square Target with Sides of 2λ Using the Method of Processing Each Source Separately and Then Combining Them in Cepstrum Space after Filtering for Three Different Permittivity Values

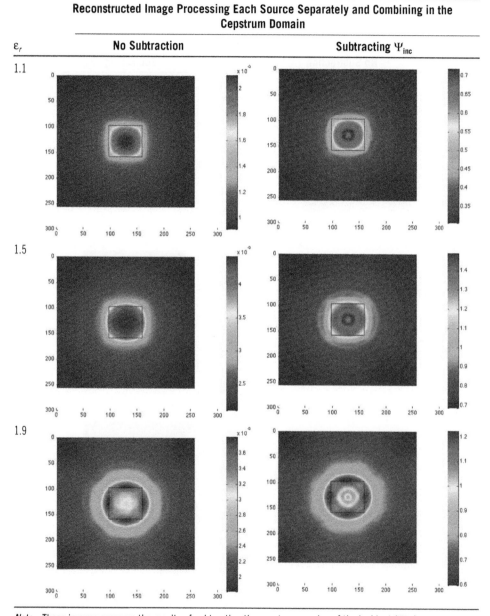

Reconstructed Image Processing Each Source Separately and Combining in the Cepstrum Domain

ε_r	No Subtraction	Subtracting Ψ_{inc}
1.1		
1.5		
1.9		

Note: These images compare the results of subtracting the cepstrum version of the incident (Ψ_{inc}).

The issue becomes critical as the spacing between the receivers equals or exceeds the wavelength of the incident field. It was also mentioned in Chapter 6 that the use of random spacing for the receivers as they are reduced in number can be very effective in dealing with this aliasing (Lustig et al., 2007) and provides a means to reduce the number of receivers needed to effectively sample the source (Candes and Romber, 2005; Candes and Wakin, 2008).

Table 10.10   A Comparison of Reconstructed Images for Square Targets with Sides of
2λ in Length Using the Method of Processing Each Source Separately and Then
Combining Them in Cepstrum Space after Filtering, Repeated for Three Different
Permittivity Values

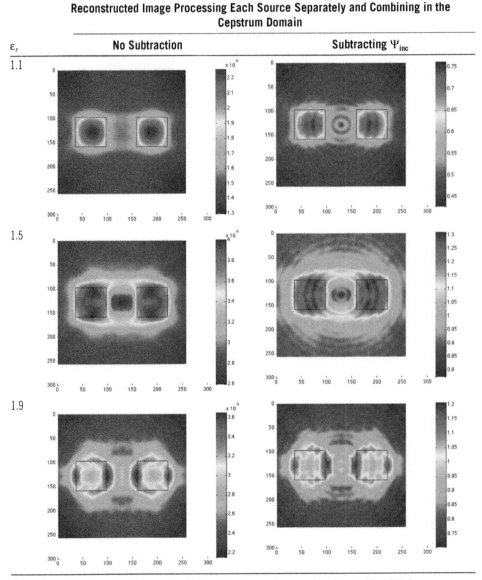

Note: These images compare the results of subtracting the cepstrum version of the incident (Ψ_{inc}).

For the model setup for the sample code in this text, the receiver spacing
from the center of the target area is 760 mm. It was observed at the output of
the cepstrum filtering method earlier that aliasing started to appear once the
number of receivers dropped below 90 for an incident frequency of 5 GHz.
The wavelength of the incident frequency of 5 GHz is 60 mm. In general, the
spacing between equispaced receivers in 2-D is calculated by dividing the
circumference by the number of the receivers:

Table 10.11   A Comparison of Reconstructed Images for Circular Targets with
Radius λ Using the Method of Processing Each Source Separately and Then
Combining Them in Cepstrum Space after Filtering for Three Different Permittivity
Values

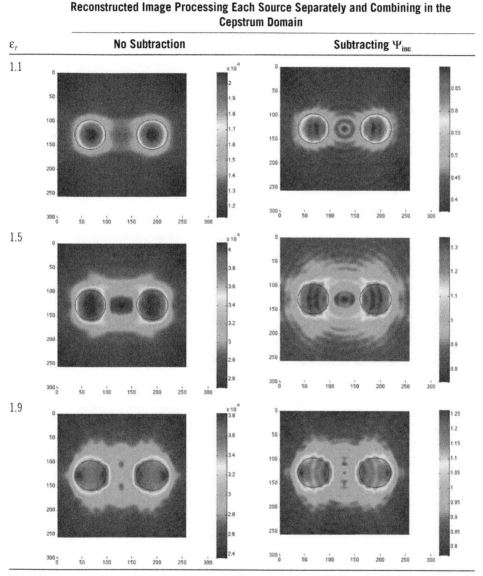

Note: These images compare the results of subtracting the cepstrum version of the incident (Ψ_{inc}).

$$\text{Spacing} = \frac{\text{Circumference}}{\text{\# of Receivers}} = \frac{2\pi r}{90} \tag{10.4}$$

For the case mentioned above, the spacing is 53 mm. This is still smaller than
the wavelength, which corresponds to the highest spatial frequency, but as the
receivers are further reduced in number, this spacing increases to be greater

Table 10.12 A Comparison of Reconstructed Images for Triangular Targets with Dimensions of 3λ Using the Method of Processing Each Source Separately and Then Combining Them in Cepstrum Space after Filtering for Three Different Permittivity Values

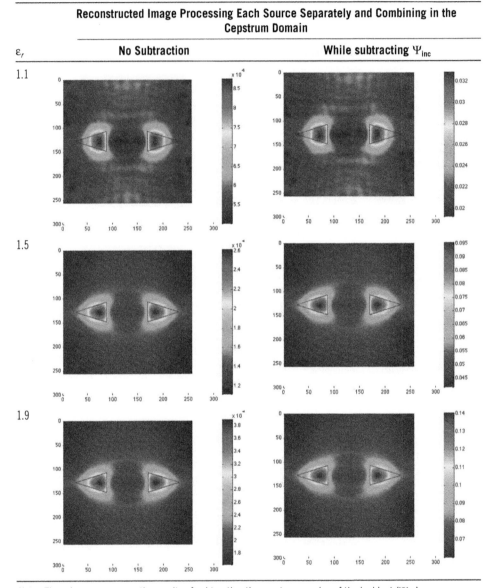

Reconstructed Image Processing Each Source Separately and Combining in the Cepstrum Domain

ε_r	No Subtraction	While subtracting Ψ_{inc}
1.1		
1.5		
1.9		

Note: These images compare the results of subtracting the cepstrum version of the incident (Ψ_{inc}).

than the wavelength, hence the increased likelihood of aliasing. For the frequency ranges used in this text, the aliasing threshold can be determined rather straightforwardly using the above relationship in Equation 10.4 and comparing it to the wavelength of the incident frequency in use. Table 10.13 shows a comparison of the number of receivers, their associated receiver spacing, and wavelengths of incident frequencies and the aliasing threshold for each.

Table 10.13 Comparison of Receiver Spacing and Incident Frequency Wavelength to Illustrate the Aliasing Threshold Limits for Each Scenario

# of Receivers	Spacing (m)	λ_{inc} (m)	F_{inc} (GHz)
360	0.013	0.030	10
180	0.027	0.033	9
		0.038	8
120	0.040	0.043	7
		0.050	6
90	0.053	0.060	5
72	0.066	0.075	4
60	0.080	0.100	3
45	0.106		
40	0.119		
36	0.133	0.150	2
30	0.159	0.300	1
24	0.199		
20	0.239		
18	0.265		
15	0.318		
12	0.398		
10	0.478		

Using the data from Table 10.13, simulations were run for each incident frequency in the range of the aliasing threshold limit predicted using the relationship described above. The results of this experiment are shown in Table 10.14 and clearly show the accuracy of predicting when aliasing will occur. Tables 10.15 through 10.18 also show examples of how applying the random receiver spacing method introduced in Chapter 6 addresses this aliasing for the images processed using the cepstrum method for a circle using various incident frequencies. In addition Tables 10.19 through 10.22 show anti-aliasing results for various target types demonstrating that the result of applying this technique to undersampled images is quite favorable in general regardless of the type of target being imaged.

Table 10.14 A Comparison of Reconstructed Images for a Circular Target with Radius λ Using the Born Approximation, and Processing Sources Independently and Recombining in Cepstrum Space

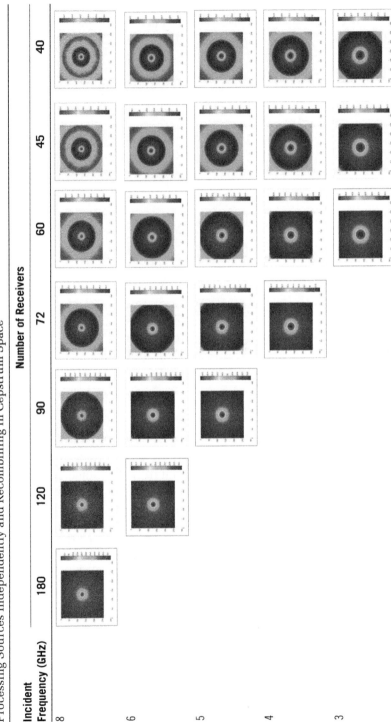

Note: Thirty-six sources are used with varying numbers of receivers equally and spaced to show the effects of aliasing after the number of receivers drops below the aliasing limit. Green indicates above the aliasing limit, yellow indicates at the threshold of the aliasing limit, and red indicates below the aliasing limit. Green square, receiver spacing less than lambda; yellow square, receiver spacing equal to or slightly greater than lambda; red square, receiver spacing greater than lambda.

Table 10.15   A Comparison of Reconstructed Images for a Circular Target with Radius λ for an Incident Frequency of 2 GHz Using the Born Approximation, and Processing Sources Independently

Freq = 2 GHz, λ = 150 mm, Radius = 150 mm = λ, Sources = 36, Resolution = 256 × 256, Scale = 1m

	Combined Ewald Circles	Born	Image Space	Cepstrum Space
ε_r = 1.5, Receivers = 36, Noise BW = 0				
ε_r = 1.5, Receivers = 18, Noise BW = 0				
ε_r = 1.5, Receivers = 18, Noise BW = 16				
ε_r = 1.5, Receivers = 18, Noise BW = 20				

Note: Thirty-six sources are used with varying numbers of receivers equally and randomly spaced to show the effects of the anti-aliasing technique of random receiver spacing.

Table 10.16 A Comparison of Reconstructed Images for a Circular Target with Radius λ for an Incident Frequency of 5 GHz Using the Born Approximation, and Processing Sources Independently

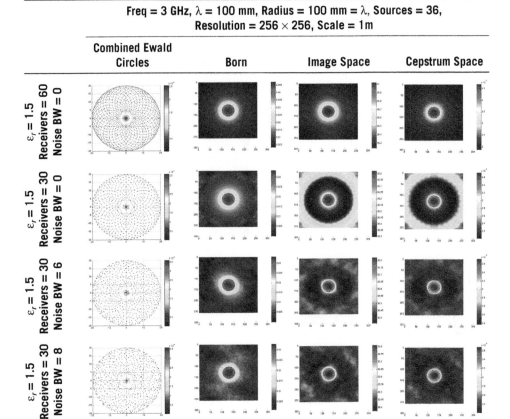

	Combined Ewald Circles	Born	Image Space	Cepstrum Space

Freq = 3 GHz, λ = 100 mm, Radius = 100 mm = λ, Sources = 36, Resolution = 256 × 256, Scale = 1m

$\varepsilon_r = 1.5$, Receivers = 60, Noise BW = 0

$\varepsilon_r = 1.5$, Receivers = 30, Noise BW = 0

$\varepsilon_r = 1.5$, Receivers = 30, Noise BW = 6

$\varepsilon_r = 1.5$, Receivers = 30, Noise BW = 8

Note: Thirty-six sources are used with varying numbers of receivers equally and randomly spaced to show the effects of the anti-aliasing technique of random receiver spacing.

Table 10.17 A Comparison of Reconstructed Images for a Circular Target with Radius λ for an Incident Frequency of 8 GHz Using the Born Approximation, and Processing Sources Independently

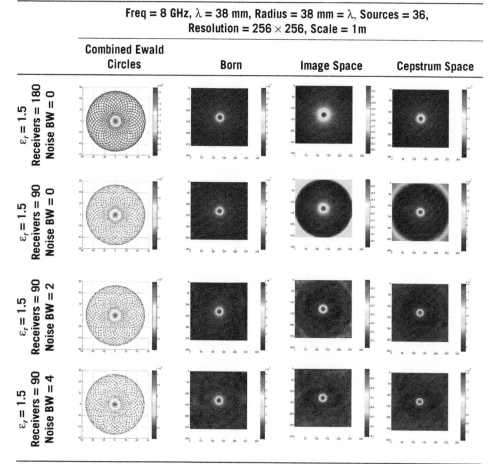

	Combined Ewald Circles	Born	Image Space	Cepstrum Space

Freq = 8 GHz, λ = 38 mm, Radius = 38 mm = λ, Sources = 36, Resolution = 256 × 256, Scale = 1 m

ε_r = 1.5, Receivers = 180, Noise BW = 0

ε_r = 1.5, Receivers = 90, Noise BW = 0

ε_r = 1.5, Receivers = 90, Noise BW = 2

ε_r = 1.5, Receivers = 90, Noise BW = 4

Note: Thirty-six sources are used with varying numbers of receivers equally and randomly spaced to show the effects of the anti-aliasing technique of random receiver spacing.

Table 10.18   A Comparison of Reconstructed Images for a Circular Target with Radius 2λ and Increased Permittivity Using the Born Approximation, and Processing Sources Independently and Recombining in Image Space, and Cepstrum Space

Freq = 5 GHz, λ = 60 mm, Radius = 120 mm = 2 λ, Sources = 36, Resolution = 256 × 256, Scale = .5 m

	Combined Ewald Circles	Born	Image Space	Cepstrum Space
$\varepsilon_r = 1.5$ Receivers = 360 Noise BW = 0				
$\varepsilon_r = 1.5$ Receivers = 30 Noise BW = 0				
$\varepsilon_r = 1.5$ Receivers = 30 Noise BW = 12				
$\varepsilon_r = 1.5$ Receivers = 36 Noise BW = 10				

Note: Thirty-six sources are used with varying numbers of receivers equally and randomly spaced to show the effects of aliasing and anti-aliasing techniques.

Table 10.19 A Comparison of Reconstructed Images for a Square Target with Sides 2λ Using the Born Approximation, and Processing Sources Independently and Recombining in Images Space, and Cepstrum Space

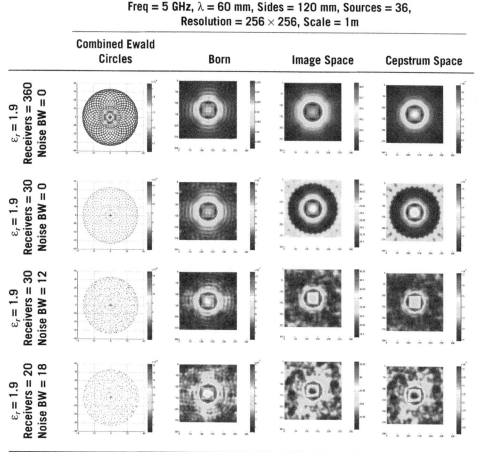

Freq = 5 GHz, λ = 60 mm, Sides = 120 mm, Sources = 36, Resolution = 256 × 256, Scale = 1m

	Combined Ewald Circles	Born	Image Space	Cepstrum Space
ε_r = 1.9, Receivers = 360, Noise BW = 0				
ε_r = 1.9, Receivers = 30, Noise BW = 0				
ε_r = 1.9, Receivers = 30, Noise BW = 12				
ε_r = 1.9, Receivers = 20, Noise BW = 18				

Note: Thirty-six sources are used with varying numbers of receivers equally and randomly spaced to show the effects of aliasing and anti-aliasing techniques.

Table 10.20   A Comparison of Reconstructed Images for a 2 Circle Targets Each with Radius λ Using the Born Approximation, and Processing Sources Independently and Recombining in Images Space, and Cepstrum Space

Freq = 5 GHz, λ = 60 mm, Radius = 60 mm = 1 λ, Sources = 36, Resolution = 256 × 256, Scale = .5 m

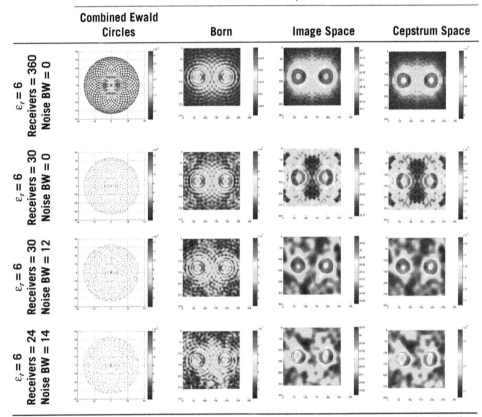

	Combined Ewald Circles	Born	Image Space	Cepstrum Space
$\varepsilon_r = 6$ Receivers = 360 Noise BW = 0				
$\varepsilon_r = 6$ Receivers = 30 Noise BW = 0				
$\varepsilon_r = 6$ Receivers = 30 Noise BW = 12				
$\varepsilon_r = 6$ Receivers = 24 Noise BW = 14				

Note: Thirty-six sources are used with varying numbers of receivers equally and randomly spaced to show the effects of aliasing and anti-aliasing techniques.

Table 10.21 A Comparison of Reconstructed Images for Two Square Targets Each with Sides 2λ Using the Born Approximation, and Processing Sources Independently and Recombining in Images Space, and Cepstrum Space

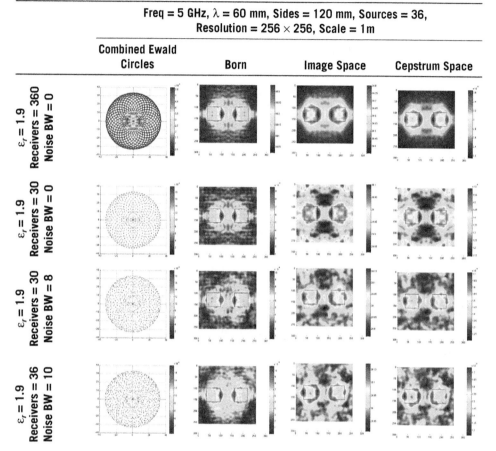

	Combined Ewald Circles	Born	Image Space	Cepstrum Space

Freq = 5 GHz, λ = 60 mm, Sides = 120 mm, Sources = 36, Resolution = 256 × 256, Scale = 1 m

$\varepsilon_r = 1.9$, Receivers = 360, Noise BW = 0

$\varepsilon_r = 1.9$, Receivers = 30, Noise BW = 0

$\varepsilon_r = 1.9$, Receivers = 30, Noise BW = 8

$\varepsilon_r = 1.9$, Receivers = 36, Noise BW = 10

Note: Thirty-six sources are used with varying numbers of receivers equally and randomly spaced to show the effects of aliasing and anti-aliasing techniques.

Table 10.22 A Comparison of Reconstructed Images for Two Triangle Targets Each with Height 3λ Using the Born Approximation, and Processing Sources Independently and Recombining in Images Space, and Cepstrum Space

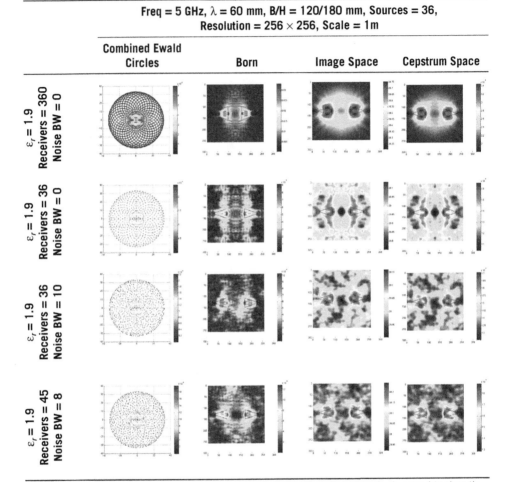

	Freq = 5 GHz, λ = 60 mm, B/H = 120/180 mm, Sources = 36, Resolution = 256 × 256, Scale = 1m			
	Combined Ewald Circles	**Born**	**Image Space**	**Cepstrum Space**
ε_r = 1.9 Receivers = 360 Noise BW = 0	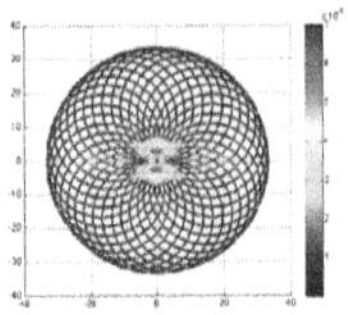	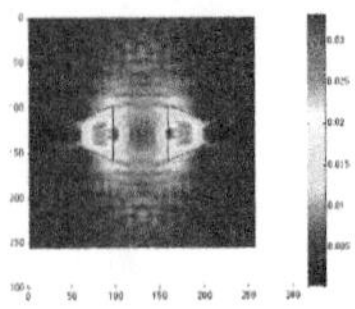	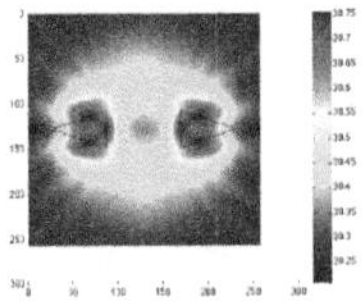	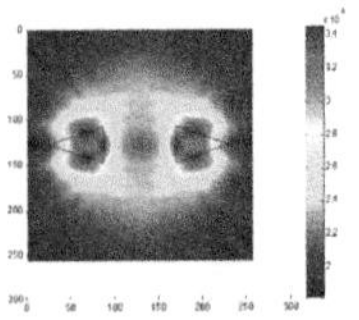
ε_r = 1.9 Receivers = 36 Noise BW = 0	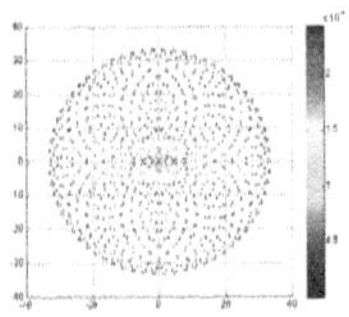	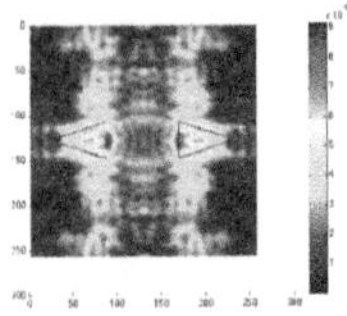	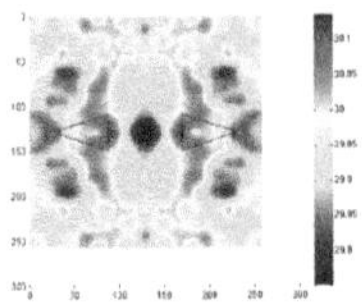	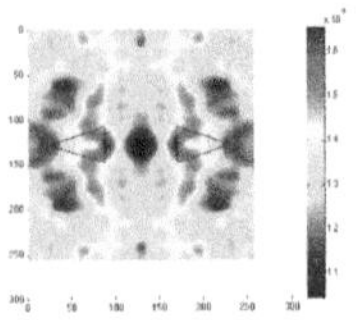
ε_r = 1.9 Receivers = 36 Noise BW = 10	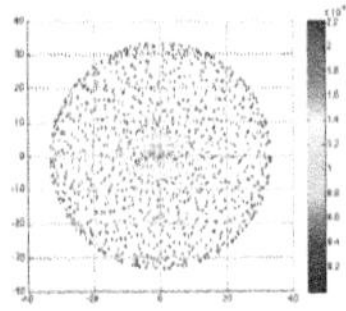	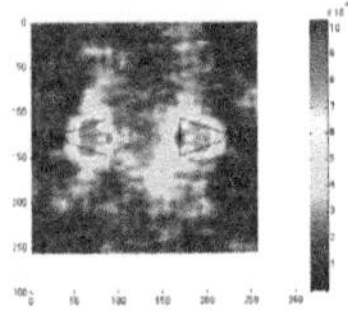	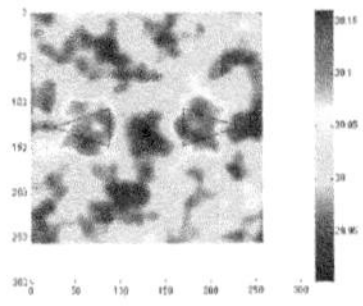	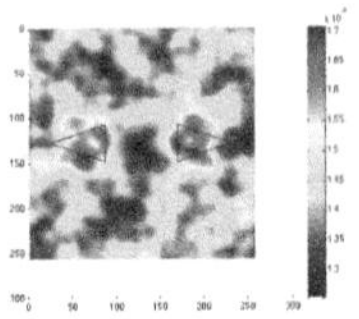
ε_r = 1.9 Receivers = 45 Noise BW = 8	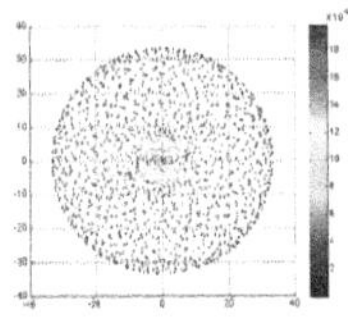	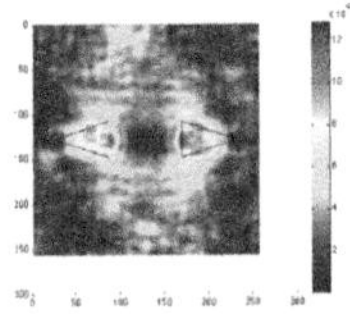	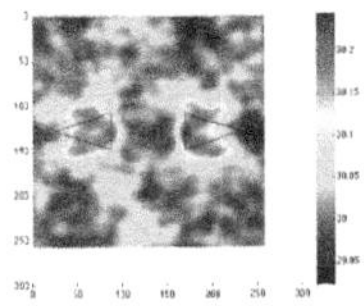	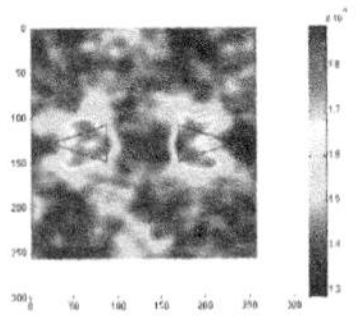

Note: Thirty-six sources are used with varying numbers of receivers equally and randomly spaced to show the effects of aliasing and anti-aliasing techniques.

REFERENCES

Candes, E. J. and Romber, J. (2005, January 25). Practical signal recovery from random projections.SPIE Vol. 5674, Computational Imaging III, pp. 76–86.

Candes, E. J. and Wakin, M. B. (2008, March). An introduction to compressive sampling. *IEEE Signal Processing Magazine*, 25(2), 21–30.

Lustig, M., Donoho, D., and Pauly, J. M. 2007. Sparse MRI: The application of compressed sensing for Rapid MR imaging. *Magnetic Resonance In Medicine*, 58(6), 1182–1195.

Shahid, U. 2009. *Signal Processing Based Method for Solving Inverse Scattering Problems.* PhD Dissertation, Optics. University of North Carolina at Charlotte, Charlotte: UMI/ProQuest LLC.

Eleven

Advanced Topics in Inverse Imaging

11.1 PRACTICAL STEPS FOR IMAGING STRONG SCATTERERS

Up to this point in this book, a gradual and historical review has been presented on approaches and advances in imaging from scattered electromagnetic fields. In the midst of this review, some very recent and significant techniques and observations have been presented that show great advancement in this field in the recent years. To demonstrate these advancements and techniques, a brief review and comparison will now be presented.

The first significant technique that was presented was that of the capability of creating a virtual test setup in a finite element modeling software such as COMSOL® to model scattered field data from a representative family of known targets. It was then shown how this data could be imported into MATLAB® for processing and investigating various algorithms and methods. It was demonstrated that this task can be successfully completed and that the results from using this structured approach can be used to produce results that are comparable to the results from experimental data when adopting the exact same experimental conditions. This new method, now validated, proves to be indispensible in producing data for many different scattering conditions that are essential in addressing the other objectives. With the use of this method and the data that it produces, existing methods and algorithms can continue to be studied and evaluated in a way that would be virtually impossible in a laboratory setting. This algorithm should prove to be very useful in future research dealing with other inverse problems, scattering, and imaging studies.

Using methods much like the one described above, data can now be generated that will allow the Born approximation and its corresponding images to be examined with great detail. In particular, the concept of the number of degrees of freedom associated with a scattering or imaging experiment that had been suggested in previous research (Miller, 2007) was put to the test. The concept of degrees of freedom for 2-D imaging from scattered field data in terms of the number of sources and receivers was studied for numerous conditions and targets. As demonstrated previously, this measure of information transfer in a scattering experiment in relation to the use of the Born approximation does seem to be valid. The degrees of freedom for each simulated experiment can be calculated, and for each case this number can be regarded as a threshold identifying the point at which the reconstructed images begin to achieve a "steady state" in their appearance. As also seen previously, this is not to say that the images at this point are "good" or that they model the original target exactly, since this depends on the extent of the multiple scattering, but that these reconstructions did not appear to show any significant improvement when using additional sources or receivers beyond this point.

The question of the quality or the appearance of each reconstructed image is altogether another issue.

The two factors of the number of degrees of freedom and the effects of resonance in determining the quality of a Born reconstructed image are extremely important. This now gives a new and possibly more meaningful set of criteria for determining or predicting when it is satisfactory to implement and expect to be able to successfully interpret an image based on the Born approximation and on methods related to Born approximation. The previous criterion of $|kV(r)a| \ll 1$ only claims to predict whether a target is a weak scatterer or not, which can be terribly subjective and inconsistent in predicting the performance of image reconstruction. The new criterion gives a much more definitive predictor since the minimum number of degrees of freedom can be calculated precisely if n_{max} is known, and the resonances might be determined or even identified for any given target.

Finally, several new techniques and algorithms were demonstrated and examined that so far have proven to be very effective. The first technique that was examined was that of the implementation of the code applying the cepstrum filtering method on an individual source basis followed by a combination of these processed data in either the image domain or the cepstrum domain. This is in lieu of an earlier approach that simply applies the cepstrum filtering method to the Born approximation composite image obtained from the combined sources. As demonstrated previously, this modified method shows a definite improvement in the appearance of the image in relation to the boundaries of the target compared to previous methods. More specifically, the procedure that combined the sources, after processing, in the cepstrum domain showed the most promise in terms of more accurate image boundary definitions. The only difficulty with these methods initially was the quantitative accuracy, in that the scale of magnitude for the improved reconstructed images did not seem to be reasonable. This issue is addressed later.

One lingering question from the cepstral filtering method above was that of the type of optimum filter to be used in the cepstrum domain. Through a series of tests shown earlier, it was determined that the optimum filter to be used is a Gaussian type filter centered at the origin using a σ value equal to 10 (for the specific conditions of the set of targets considered here). Another striking discovery was that the absolute peak value of the filter plays a critical role in determining the magnitude of the reconstructed image produced by the new cepstrum method described above. In particular, it was shown that the magnitude of the output of the method that processes each source individually and recombined them in the cepstrum domain was inversely proportional to the magnitude of the peak of the Gaussian filter. After further examination, it was shown that the peak value of the Gaussian filter needed to be scaled by a factor of $2/(3N_s)$, where N_s is the number of sources, to let the magnitude of the reconstructed images be of the same range of the magnitude of the Born reconstructed image for the same data. It is not known or understood at this time why this scaling factor is needed. It could be a result of applying 3-D techniques to a 2-D problem or some variance of this. In addition to these experiments on the filter characteristics in cepstrum space, a new approach was derived and evaluated for potential benefits as well. That new approach involved subtracting a weighted cepstrum of the incident field during the processing of the sources in cepstrum space. As explained earlier, this is done in an attempt to reduce or eliminate the $\langle\Psi\rangle/R'$ term in the cepstrum domain representation of the signal. The results of applying this additional "filtering" step proved to be very promising in that it appears to improve the scale of the magnitude of the reconstructed image resulting from the "combined in cepstrum space" method. In fact, as shown in the previous results, the peak of the scale is very near the "correct" value for the permittivity (or better yet, to the index of refraction) of the original target.

If we now take all of the improvements mentioned above, that is, processing sources independently, optimized filtering, and subtracting the incident field in the cepstrum domain, we can apply these new and improved methods to a more complex target set and compare its results to those obtained by using previous methods to observe the progression of improvements. This was done for a target set consisting of a circle with radius of λ, a square with sides of 2λ, and a triangle with a base of 2λ and a height of 3λ. This concluding test was run three times with the targets having a permittivity of 1.1, 1.5, and 1.9 respectively to observe how the methods performed in relation to each other. The results of this set of demonstration experiments are shown in Figures 11.1 through 11.3 below. The results shown in Figure 11.1 are extremely

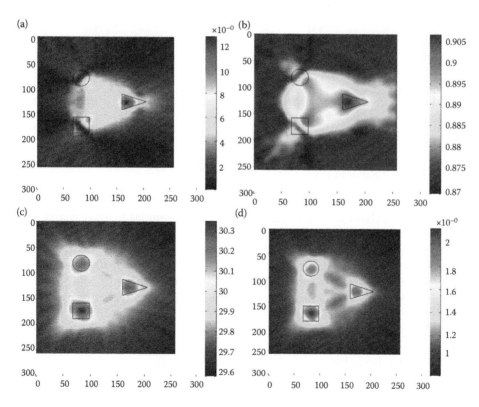

Figure 11.1 Comparison of reconstructed images from various method outputs for a target set consisting of a circle of radius λ, a square with sides 2λ, and a triangle with base 2λ and height 3λ. All targets have a permittivity of 1.1. The outputs shown above are from (a) Born approximation, (b) cepstrum of image in (a) using algorithm developed in Shahid (2009), (c) cepstrum filtering of individual sources recombined in image space, and (d) cepstrum filtering of individual sources that are recombined in image space and have a cepstrum version Ψ_{in} subtracted in cepstrum space.

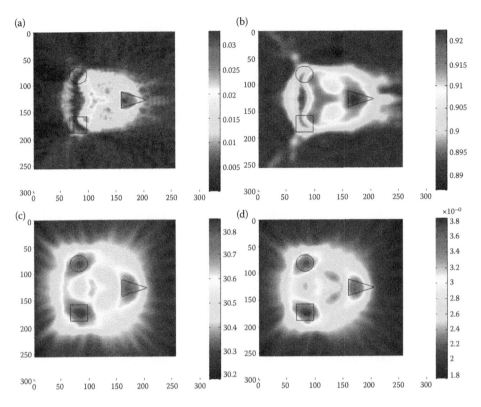

Figure 11.2 Comparison of reconstructed images from various method outputs for a target set consisting of a circle with radius of λ, a square with sides of 2λ, and a triangle with base of 2λ and height of 3λ. All targets have a permittivity of 1.5. The outputs shown above are from (a) Born approximation, (b) cepstrum of image in (a) using algorithm developed in Shahid (2009), (c) cepstrum filtering of individual sources that are recombined in image space, and (d) cepstrum filtering of individual sources that are recombined in image space and have a cepstrum version Ψ_{in} subtracted in cepstrum space.

encouraging in that the improvement seen going from (a) to (d) is really quite surprising in relation to the target boundaries. The output for method (d) clearly shows that there are three distinct targets and makes valid attempts to show the extent of these boundaries. The scale of the magnitude for (d) is also very close in range for the index of refraction for the targets of 1.04.

The results shown in Figure 11.2 are encouraging as well, though maybe not quite as striking as the results shown in Figure 11.1. Again, improvement is seen going from (a) to (d) in relation to the target boundaries. The output for method (d) clearly shows again that there are three distinct targets and makes a fair attempt to show the extent of these boundaries. The scale of the magnitude for (d) is again very close, actually almost exact in the range for the index of refraction for the targets, which has an index value of 1.225. This is not as good an overall performance as that shown in Figure 11.1, but still very encouraging.

The results shown in Figure 11.3 indicate progressively poor overall performance as expected, but the image in (d) does show continued improvement compared to the other methods. As previously stated, improvement is seen going from method (a) to (d) especially in relation to the target boundaries. The output for method (d) does show again that there are three distinct

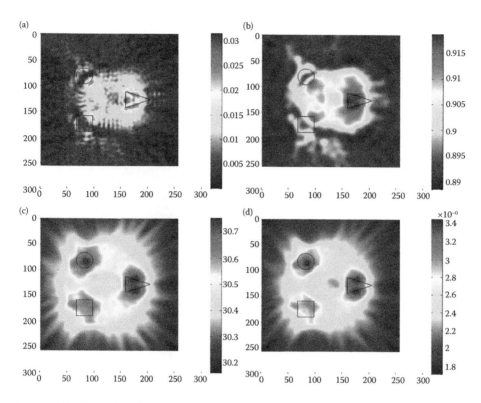

Figure 11.3 Comparison of reconstructed images from various method outputs for a target set consisting of a circle with radius of λ, a square with sides of 2λ, and a triangle with base of 2λ and height of 3λ. All targets have a permittivity of 1.9. The outputs shown above are from (a) Born approximation, (b) cepstrum of image in (a) using algorithm developed in Shahid (2009), (c) cepstrum filtering of individual sources that are recombined in image space, and (d) cepstrum filtering of individual sources that are recombined in image space and have a cepstrum version Ψ_{in} subtracted in cepstrum space.

targets, but its attempt at defining the boundaries is fairly poor. The scale of the magnitude for (d) is also very close in range for the index of refraction for the targets which has an index value of 1.378.

While these demonstrations are encouraging and show great improvement and promise, there are still two apparent immediate opportunities for future research. The first is fairly straightforward in that the issue of the scale of the new cepstrum method generated image is still not fully resolved. Great improvements have been incorporated and we have a better understanding than we have had in the past, but it is still not fully understood. Also, there appears to be some other nonlinear dependence on either the number of sources, the permittivity of the target or both. There may be other opportunities to remove or reduce the effects of the $\langle\Psi\rangle$ term in the cepstrum domain much like the ones utilized previously in this book.

11.2 AN OVERALL APPROACH TO THE DEGREES OF FREEDOM IN IMAGING

It was noted earlier in this book that, as experiments were being performed on the minimum required number of receivers in relation to the degrees of

Table 11.1 Ewald Circle Sets with Accompanying Born Reconstructed Images for a Series of Source/Receiver Combinations All Totaling 216 Receivers (or Data Points) with Random Spacing Applied to the Receiver Locations about Their Original Equispaced Locations

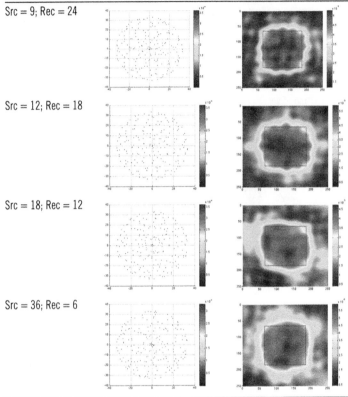

Src = 9; Rec = 24	
Src = 12; Rec = 18	
Src = 18; Rec = 12	
Src = 36; Rec = 6	

Note: Target used is a square with sides equal to 2λ and a permittivity of 1.5. Src, source; Rec, receiver.

freedom, a randomness was injected into the receiver locations to address aliasing. In this process, there came a point where it was difficult to distinguish the number of sources from the number of receivers that were actually present. It had begun to look more like as if it was a matter of the total number of receiver points (and possibly their distribution) more than how many sources there were, that is, perhaps the more important metric is the total number of data points that are evenly distributed. This is illustrated in Table 11.1. For each of the set of Ewald circles and image reconstructions, the total number of receiver or data points is the same, which is 216. Because they are randomly spaced (at least around their original locations), it is difficult to distinguish between the Ewald circle sets since the resulting images appear to be very similar in appearance. This would seem to suggest that possibly the minimum requirement is not so much a function of the number of sources and/or the number of receivers required, but more the total number of receiver or data points equally but randomly spaced to fill the k-space area. If this is the case, then it is possible that the minimum degrees of freedom may not be so much a function of either sources or receivers by themselves, but an expression involving a combination of the two.

The first intuitive response could be that it may be simply a product of the two or $N_s \times N_r$, which would simply be N^2. This could be tested quickly using the model previously used. For this scenario N_s/N_r is calculated to be 4.9, so for these conditions we could declare that $N = 5$. If in fact N^2 is the relationship, we should be able to satisfy this by setting $N_s = 6$ and $N_r = 5$. The results are shown in Figure 11.4 and compared with the case for $N = 12,960$.

Reviewing the results in Figure 11.4, it does not appear that the "overall" number of degrees of freedom has been satisfied. The combined requirement for the sources and receivers, if this is in fact the case, may be as simple as applying a Nyquist–Shannon theorem type approach which might look something like $2N_s \times 2N_r = 4N^2$. Applying this criterion to the square model used before would mean that the minimum number of degrees of freedom or data points required would be $N = 4(5)^2 = 100$. This can be achieved (or exceeded) in the square model by using 12 sources and 10 receivers randomly spaced. This too was implemented with the results shown in Figure 11.5, again along with the case for $N = 12,960$.

The image obtained from using 12 sources and 10 receivers shown in Figure 11.5b does seem to confirm that the minimum degrees of freedom have been met when compared with the image in Figure 11.5d for the case with 36

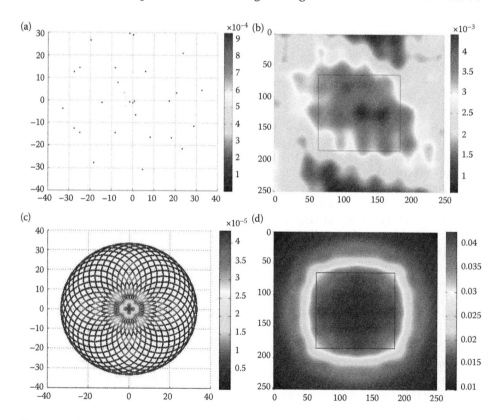

Figure 11.4   (a) Ewald circle using six sources and five receivers randomly spaced. (b) Resulting reconstructed image from (a). (c) Ewald circle using 36 source and 360 receivers equally spaced for comparison with (a) and (b). (d) Resulting image from (c).

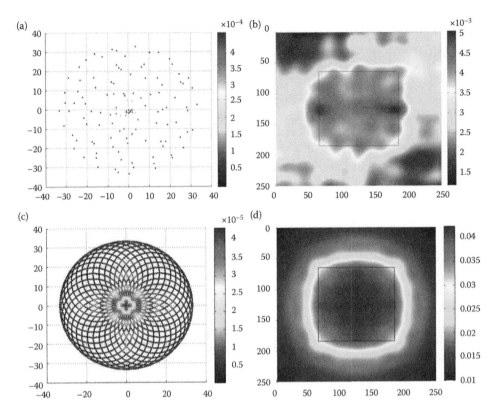

Figure 11.5 (a) Ewald circle using 12 sources and 10 receivers randomly spaced. (b) Resulting reconstructed image from (a). (c) Ewald circle using 36 source and 360 receivers equally spaced for comparison with (a) and (b). (d) Resulting image from (c).

sources and 360 receivers. It is entirely possible that the minimum has been met in this case. There simply is not enough evidence at this point to say conclusively if this describes the overall degrees of freedom relationship.

Having now observed the evidence that there does seem to be an overall degrees of freedom criterion, it seems appropriate not to test this idea a bit more thoroughly to see if this concept holds true and if so, whether there appears to be a consistent threshold that we can identify and possibly even quantify as an overall degrees of freedom criterion. To do this, we will examine two different targets over a range of permittivites for the scenarios of the number of randomly but equally spaced data points that correspond to $N^2/2$, N^2, $2N^2$ and $4N^2$. These images are then compared to the maximum degrees of freedom case for this experiment which is $N = 12,960$ for comparison. The results from testing circular targets are shown in Table 11.2. It is clear from these images that, for the cases of $N^2/2$ and N^2 total data points, the degrees of freedom are clearly not satisfied for any of the permittivity cases. The images are starting to take shape for the $2N^2$ case for all permittivity values, but still appear as if they have not quite satisfied the overall number of degrees of freedom. Once the number of data points reaches the $4N^2$ case, it is evident that the image is fairly well formed for all permittivity ranges. This strongly suggests that the overall degrees of freedom criterion have been satisfied at this point for all permittivities for this class of target.

Table 11.2   Family of Reconstructed Images for a Circle with a
Radius of 1λ for Varying Permittivity Values and Varying
Number of Randomly Spaced Data Points in k-Space

	$\varepsilon_r = 1.1; N = 3.3$	$\varepsilon_r = 1.5; N = 3.8$	$\varepsilon_r = 1.9; N = 4.3$
NN/2			
	8	8	15
NN			
	16	16	30
2NN			
	32	32	48
4NN			
	72	72	108
	12,960	12,960	12,960

Next, we look at the results from the square targets shown in Table 11.3. It
again is clear that, for the cases of $N^2/2$ and N^2 total data points, the degrees
of freedom are clearly not satisfied for any of the permittivity cases. Likewise,
as before, the images are starting to take shape for the $2N^2$ case for all permittivity values, but still appear as if they have not quite satisfied the overall
number of degrees of freedom. Finally, again as before, once the number of
data points reaches the $4N^2$ case it is evident for all permittivity ranges that
the image is fairly well formed. This even strengthens the idea that there is
an overall degrees of freedom criterion and that this limit is in the neighbor-
hood of $4N^2$. This needs more exhaustive testing before it can be declared as a
theorem, but it surely presents a strong case for the relationships suggested for
overall degrees of freedom for imaging while using these techniques.

11.3 CONCLUSION

The problem of imaging from scattered fields is at least 100 years old and continues to provide challenges, both theoretical and experimental, across liter-
ally a broad spectrum of probing waves. The interaction of waves with matter,
be they acoustic or electromagnetic, is complex; even the direct problem, that

Table 11.3 Family of Reconstructed Images for a Square with Side Lengths of 2λ for Varying Permittivity Values and Varying Number of Randomly Spaced Data Points in k-Space (Square with Sides = 2λ)

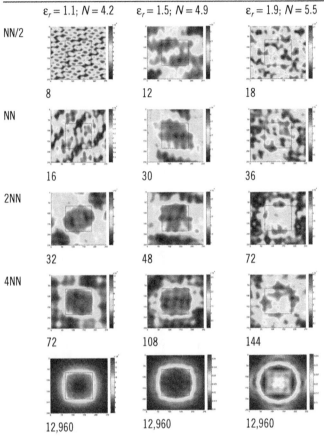

	$\varepsilon_r = 1.1$; $N = 4.2$	$\varepsilon_r = 1.5$; $N = 4.9$	$\varepsilon_r = 1.9$; $N = 5.5$
NN/2	8	12	18
NN	16	30	36
2NN	32	48	72
4NN	72	108	144
	12,960	12,960	12,960

is, of predicting the scattered field from a known scattering object can be difficult. We have focused here primarily on the scattering of electromagnetic waves and provided an overview of those interactions and, we hope, a historical perspective on how concepts and models have developed over the years. The most tractable modeling approach is clearly to linearize the problem. More precisely, assuming both linearity and time or space invariance of these governing models allows elegant Fourier-based techniques to be applied. We cite, for example, the huge literature on Fourier optics, a subset of physical optics, and the widely used first Born approximation. A Fourier relationship between a scattering function and the measured field allows a wealth of methods for image recovery and enhancement to be brought to bear on the problem. We have discussed how to address the limited data problem, resolution enhancement, and phase information retrieval based on this underlying model.

In practice, scattering is rarely sufficiently weak or involves such a sufficiently slowly varying scattering parameter (Rytov approximation) for these weakly scattering models to apply. Much effort has gone into trying to better understand the consequences of using them when they are not strictly valid, and of course far more effort has gone into developing methods that attempt to directly address the strong scattering problem. We have only touched briefly on most of these alternate methods here, because they require a deep theoretical treatment that is beyond the scope of this book. Also, they often require considerable prior knowledge and/or sophisticated pre- and postprocessing in order to guarantee convergence to an unbiased solution. What we presented here was an alternative and one-step nonlinear inverse scattering method, which returns to the strategy of trying to linearize a blatantly nonlinear problem. By considering the integral equation of scattering for the strongly scattering situation, we model the Born approximation reconstruction as an image corrupted by multiplicative band-limited noise (i.e., the different realizations of the total field within the scattering volume). This allowed us to apply cepstral filtering, that is, linear filtering techniques drawn from Fourier-based theories to the logarithm of the recovered "image." The steps involved in processing this algorithm were described and several important conclusions were drawn.

The first was that this method does indeed work quite effectively but only if there were sufficient data to work with. Based on a more fundamental analysis of a scattering experiment, one can identify the number of degrees of freedom of that system, rather like viewing it as an information channel. In order to recover an image in which one can have some confidence, the scattering geometry and maximum values for the scattering parameters (i.e., size and index) allow us to define a channel capacity or the number of degrees of freedom associated with that imaging experiment. Only when a sufficient number of data have been collected, for example, by either varying the source and/or the receiver locations, can one expect to obtain a reliable image, the problem being underdetermined otherwise. Being Fourier based, incorporation of prior knowledge can alleviate limited data problems, but one must always be careful about how this is done in order not to bias the recovered image. Also, we noted that the precise form of the cepstral filter used is not that critical, and more work could be done for any given problem to optimize this step, depending on the scattering object of interest.

It was also interesting to note that one cannot ignore resonant effects when illuminating scattering objects. The size, shape, and material properties can lead to energy storage in the scatterer, and there is always the fundamental issue of addressing possible nonscattering structures leading to significant ambiguities. We probed the simplest of resonant structures with this concern in mind, namely, dielectric spherical scatterers, since there is an exact solution available for spheres whose size is of the order the wavelength. The solution is expressed in terms of an infinite series, but a truncated series can provide very good estimates for the scattered field from various spherical objects. We found that at resonances when the internal field in the scatterer is at its maximum, the cepstral-based recovery of the image of the spherical shape improved. This indicates that the more the complexity of the scattered field inside the scattering object, the better this inversion method works. For images of dielectric spheres based on this Mie scattering model, we found that reconstructions (Born and cepstral) were better close to a Mie resonance, that is, at a relatively high Q situation.

Also, in the presence of sufficient data as dictated by the number of degrees of freedom, the recovered image of these objects reflected their scattering cross sections rather than their actual physical dimensions. Their actual sizes can be deduced from size-sensitive estimation techniques like the PDFT that was described in this book. Another important observation regarding the resolution of the resulting image of a strongly scattering object should be made. Since, as we described, sub-wavelength-sized scattering features in an object can generate evanescent waves, the further scattering of these waves back in to propagating waves can lead to subtle changes in far fields that are measured. It is not unusual to see apparent super-resolved features in the images of strongly scattering objects for this reason.

Clearly much more needs to be done to advance the overall state of the art in imaging from scattered fields. The algorithm described here is a step in that direction. It would be remiss not to end with another application of growing importance for which this and related inverse scattering algorithms can be very useful. If one has a method to recover an image of an index or permittivity distribution from a sufficiently large number of scattered field measurements, then one can apply the same steps to synthesize objects that may not exist. By specifying scattered fields as a function of angle, wavelength, incident field direction, etc., these fields can be inverted to create an object distribution that can be used to design and create new objects and materials. Constraints on the index modulation or form factor of the object can be imposed using a method like the PDFT during the reconstruction (i.e., synthesis) step. We have applied this approach with some success to the design of low observable objects and object shape shifting covers. In other words, we can use the imaging method described to design structures that are very difficult to image because they scatter very little in selected directions, or scatter like some structure having a different appearance.

Recent interest in engineered materials or meta-materials and their role in the interaction of electromagnetic waves can benefit from the inverse methods presented here. For example, close to a strong resonance, a Mie resonance associated with a relatively high Q situation, one can expect more extreme effective values of refractive index (or permittivity or permeability). The first and dominant Mie resonance for a dielectric particle is associated with a magnetic and not electrical resonance. This has led to recent increasing interest by the meta-material community in getting rid of lossy metals in meta-materials and achieving negative permeability and negative permittivity to achieve a negative index just by using two different species of dielectric spheres in their meta-materials. An interesting and important problem worth studying would be to see if one could design an improved meta-material using inverse scattering methods based on these observations.

REFERENCES

Miller, D. A. 2007. Fundamental limit for optical components. *Journal of the Optical Society of America B*, 24(10), A1–A18.

Shahid, U. 2009. *Signal Processing Based Method for Solving Inverse Scattering Problems*. PhD Dissertation, Optics. University of North Carolina at Charlotte, Charlotte: UMI/ProQuest LLC.

IV

APPENDICES

VI

APPENDICES

Appendix A: Review of Fourier Analysis

BACKGROUND TO THE FOURIER TRANSFORM

The Fourier transform in 1-D, with the notation for k-space, is written as

$$F(k_x) = \int\limits_{-\infty}^{+\infty} f(x)e^{i(k_x x)}dx \qquad (A.1)$$

The Fourier transform in 2-D k-space is

$$F(k_x, k_y) = \int\limits_{-\infty}^{+\infty}\int\limits_{-\infty}^{+\infty} f(x,y)e^{i(k_x x + k_y y)}dx\,dy = \int\limits_{-\infty}^{+\infty}\left\{\int\limits_{-\infty}^{+\infty} f(x,y)dy\right\}e^{i(k_x x)}dx \qquad (A.2)$$

For the 1-D example, we can write this as

$$F(k_x) = \int\limits_{-\infty}^{+\infty} f(x)(\cos(k_x x) + i\,\sin(k_x x))dx \qquad (A.3)$$

This can be interpreted as the multiplication of $f(x)$ by a cosine and a sine function followed by integration over all of x. Another description of this is a "projection" of a cosine or sine onto $f(x)$. This operation provides a measure of the relative weight of each specific (sine and cosine) spatial frequency that is required to reconstruct $f(x)$ from this particular basis or expansion set of functions representing $f(x)$, that is, a basis set that happens to be sines and cosines. The same terminology applies if we were to represent $f(x)$ using a different basis or set of functions such as those discussed in the chapter on the PDFT (Chapter 7 and Appendix C). $F(k_x)$ is a complex number and the Fourier coefficient for the spatial frequency k_x.

The case when $k_x = 0$, this gives $F(0) = \int_{-\infty}^{+\infty} f(x)dx$ which is a measure of the area under $f(x)$ or, in terms of the Fourier transform, is the zero frequency coefficient which is a measure of the DC level in $f(x)$. The inverse Fourier transform is written as $f(x) = \int_{-\infty}^{+\infty} F(k_x)e^{-i(k_x x)}dk_x = F^{-1}\{F\}$. Clearly, we have $f = F^{-1}F\{f\}$.

Some properties of the Fourier transform are:

Linearity: $F\{Af + Bg\} = AF + BG$
Similarity or scaling: $F\{f(ax)\} = [F(k_x/a)]/|a|$

THE DIRAC DELTA FUNCTION

This function is very convenient and widely used and best understood in terms of the following short derivation. If

$$f(x) = \int\limits_{-\infty}^{+\infty} F(k_x)e^{-i(k_x x)}dk_x = \int\limits_{-\infty}^{+\infty}\left[\int\limits_{-\infty}^{+\infty} f(x)e^{i(k_x x)}dx\right]e^{-i(k_x x)}dk_x \tag{A.4}$$

then this gives us the opportunity to define the delta function by writing

$$f(x) = \int\limits_{-\infty}^{+\infty} f(x')\delta(x' - x)\,dx' = \int\limits_{-\infty}^{+\infty}\left[\int\limits_{-\infty}^{+\infty} f(x)e^{i(k_x x)}dx\right]e^{-i(k_x x)}dk_x \tag{A.5}$$

leading to the identity

$$\delta(x' - x) = \int\limits_{-\infty}^{+\infty} e^{i(k_x x' - k_x x)}dk_x \tag{A.6}$$

$$F\{\delta(x)\} \Rightarrow \int\delta(x)e^{-i\omega x}dx = e^{-i\omega 0} = 1 \tag{A.7}$$

If $f = e^{-i\omega_0 t}$ then $F(\omega) = \int\limits_{-\infty}^{\infty} e^{i\omega_0 t} \times e^{-i\omega t}dt = \int\limits_{-\infty}^{\infty} e^{-i(\omega - \omega_0)t}dt = \delta(\omega - \omega_0)$

If $f = \cos\omega_0 t$ then

$$F(\omega) = \frac{1}{2}\int\limits_{-\infty}^{\infty} (e^{i\omega_0 t} + e^{-i\omega_0 t})e^{-i\omega t}dt \tag{A.8}$$

$$= \frac{1}{2}\int\limits_{-\infty}^{\infty} e^{-i(\omega - \omega_0)t}dt + \frac{1}{2}\int\limits_{-\infty}^{\infty} e^{-i(\omega + \omega_0)t}dt \tag{A.9}$$

$$= \frac{1}{2}\delta(\omega - \omega_0) + \frac{1}{2}\delta(\omega + \omega_0) \tag{A.10}$$

Examples of Some Notations

Using the similarity or scaling theorem, that is,

$$g(ax)\xrightarrow{\text{F.T.}}\frac{1}{|a|}G\left(\frac{u}{a}\right) \tag{A.11}$$

then, $\text{rect}(ax)$ is $1/a$ units wide $\xrightarrow{\text{F.T.}} 1/a$ sinc (u/a) which has equidistant real zeros at every $u/a = m\pi$, where the $\text{rect}(x)$ function is defined in Figure A.1 and the Fourier transform of this function is illustrated in Figure A.2. Fourier transform symmetries can be found in Figure A.3.

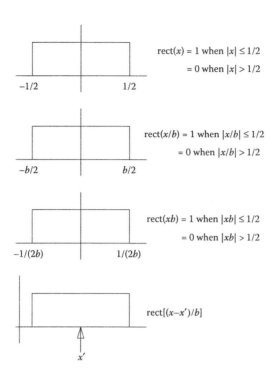

Figure A.1 Examples of rect notation.

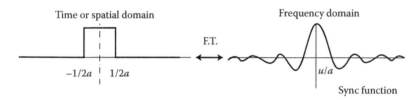

Figure A.2 Example of Fourier transform of rect function.

Convolution and Correlation

$F = F\{f\}$ and $G = F\{g\}$, then

1. $h(x) = \int\limits_{-\infty}^{\infty} f(x)g(x)\mathrm{d}x$ is projection of g on f

2. $h(x) = \int\limits_{-\infty}^{\infty} f(x')g(x - x')\mathrm{d}x'$ is convolution of g with f.

3. $h(x) = \int\limits_{-\infty}^{\infty} f(x')g(x' - x)\mathrm{d}x'$ is the correlation of g with f.

We can think of convolution as a sliding projection but notice that $g(x)$ becomes $g(-x)$ in the integral and so is mirrored about the x origin. If $H = F\{h\} = FG$, we have case (ii) in the Fourier domain. For case (iii), $H = FG^*$ where the asterisk (*) denotes complex conjugate. A graphical illustration of correlation can be found in Figures A.4 and A.5.

$$F\{h(x)\} = F\left\{\int_{-\infty}^{\infty} f(x')g(x-x')dx'\right\} = \int_{-\infty}^{+\infty} h(x)e^{i(k_x x)}dx$$

$$= \int_{-\infty}^{+\infty}\left[\int_{-\infty}^{\infty} f(x')g(x-x')dx'\right]e^{i(k_x x)}dx$$

(A.12)

Let $x - x'$ be written as y then $dy = dx$ and we can write

$$F\{h(x)\} = \int_{-\infty}^{+\infty}\left[\int_{-\infty}^{\infty} f(x')g(y)dy\right]e^{i(k_x(y+x'))}dx'$$

$$= \int_{-\infty}^{+\infty} f(x')e^{i(k_x x')}dx'\int_{-\infty}^{+\infty} g(y)e^{i(k_x y)}dy = F(k_x)G(k_x)$$

Graphical illustration of convolution:

If $g = f$, then a correlation function is termed as an autocorrelation function, and its Fourier transform is an energy spectrum (Wiener–Khinchin Theorem)

Examples of convolutions:

$$\text{triangle}(x) = \text{rect}(x) * \text{rect}(x)$$

(A.13)

$$F\{\text{triangle}(x)\} = \text{sinc}(u) \cdot \text{sinc}(u) = \text{sinc}^2(u)$$

(A.14)

$$f(x) * \delta(x - x_0) = f(x - x_0)$$

(A.15)

In the Fourier domain, these convolutions are simple products of the functions' Fourier transforms, thus

$$\text{if } f(x) = \text{rect}(bx), F(u) = \frac{1}{b}\text{sinc}\left(\frac{u}{b}\right)$$

If	Then
$h(t)$ real	$H(-u) = [H(u)]'$
$h(t)$ imaginary	$H(-u) = -[H(u)]'$
$h(t)$ even	$H(-u) = H(u)$ (even)
$h(t)$ odd	$H(-u) = -H(u)$ (odd)
$h(t)$ real and even	$H(u)$ real and even
$h(t)$ real and odd	$H(u)$ imaginary and odd
$h(t)$ imaginary and even	$H(u)$ imaginary and even
$h(t)$ imaginary and odd	$H(u)$ real and odd

Figure A.3   Fourier transform symmetries.

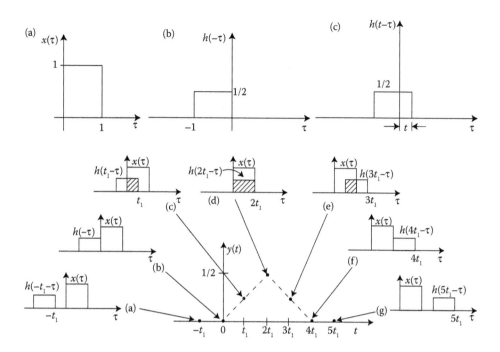

Figure A.4   Example of convolution of rect function.

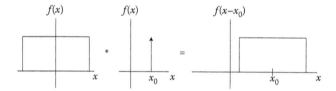

Figure A.5   Example of rect function convolved with impulse function.

$$f(x - x_0) \rightarrow F(u)e^{-ix_0u} = e^{-ix_0u}\frac{1}{b}\mathrm{sinc}\left(\frac{u}{b}\right)$$

Multiplication with a linear phase function in x or u leads to a shift of the function in the respective Fourier spaces and vice versa. We have $\delta(x \pm x_0) \xrightarrow{\text{F.T.}} e^{\pm ix_0u}$ (physically this is a "tilted" plane wave).

There are a number of useful symmetries associated with Fourier transforms.

Appendix B: The Phase Retrieval Problem

At high frequencies, that is, above a few terahertz, it is increasingly difficulty to measure the phase of a scattered field. The inverse scattering methods described here assume that one can measure both the magnitude and the phase of the scattered field. The (Fourier) phase retrieval problem is an old one with many different approaches to its solution depending on the nature of the problem (e.g., compact vs. periodic structures).

Given a 3-D transparent object, interferometry or making a hologram using the scattered field can be used to image it. Classical interferometric techniques are limited to the determination of 2-D fields since they provide the path length data for single directions of illumination. Holographic interferometry makes it possible to obtain an interferogram with multidirectional illumination and determine the optical path length of "rays" passing through the object in many different directions. There remains the problem of relating the refractive index distribution to these data, and the assumption generally has to be made that rays travelling in straight lines model the situation well. Some work on ray tracing in refracting objects has been carried out. In any such procedure there are considerable experimental difficulties, such as the uncertainty in the fringe order when refractive index gradients change sign with the object.

One of the few experimental and numerical attempts to determine the structure of an inhomogeneous scattering object was due to Carter (1983). To simplify matters a 3-D object, but one having a 1-D scattering potential, was made from two homogeneous rectangular blocks differing in refractive index by 0.222 and made sufficiently small such that the first Born approximation was valid. The complex scattered fields containing the information required to recover quantitatively the 3-D structure of this object were determined holographically, following the procedure suggested by Wolf (1970). Wolf showed that from measurements of the intensity distribution of the field emerging from an off-axis hologram, one can determine the complex field down to details of ~9 wavelengths with a reference beam angle of −30°. Carter scanned a small region of such a hologram around the propagation direction, converted optical density to effective exposure, and determined the average fringe period. These data were then used to demodulate digitally the amplitude and phase information carried by the hologram fringes. This procedure would then have to be repeated for all incident wave directions and should be repeated to reduce noise. Repeats were performed in the experiment described here for experimental details for a comparison between the Born and Rytov approximations applied to these data.

One of the main problems encountered in this approach is the difficulty in maintaining a precise reference and using a holographic technique while varying the direction of illumination. This difficulty and the amount of data required to determine the 3-D structure are clearly reduced if some assumptions about object symmetry can be made. It was concluded that since the experimental procedure is so complicated even for a simple object, further experimental or theoretical advances would be required in order to exploit fully the potential value of inverse scattering in many fields.

The most attractive outcome would be the availability of an algorithm to calculate the phase of the scattered field directly from its measured intensity distribution without the need for a reference wave. We discuss this next, and in more detail, the problem of limited Fourier data. The phase retrieval problem has been the subject of many reviews and can be divided into 1-D problems and more than 1-D problems.

In either case one looks to the analytic properties of the scattered field to indicate the range of possible phase ambiguities. In 1-D, a band-limited field can be represented by an infinite product of linear factors, each describing a zero of the field in the complex plane. Complex zero locations can be complex conjugated and leave the magnitude changed but not the phase, thus generating phase ambiguities. In more than 1-D, depending upon the object, the field may factor into any number of factors or none at all, making zeros difficult to characterize and the phase ambiguities unclear. It is generally assumed that for a function of a continuous variable the phase is unique in 2-D or 3-D, but with a finite number of samples, there is necessarily nonuniqueness.

Let us assume that a Fourier relation can be written between the scattered field and an "object function" which is of finite support. This, for example, is readily satisfied if the measurements of the scattered field are taken in the Fraunhofer or Fresnel regions.

The scattered field is a band-limited function from a compact structure, and such functions, in one or more variables, have remarkable properties. A finite Fourier transform is an entire function of exponential type and so remains analytic everywhere in the complex plane with well-defined growth properties. These properties provide a basis on which one can specify the relationship, if any, between the magnitude and phase of a band-limited function and determine whether or not the phase function is well defined at all points of interest.

By definition, a band-limited function, $F(x)$, may be expressed as the finite Fourier transform of an object function, $f(t)$, as follows:

$$F(x) = \int_a^b f(t)e^{ixt}\mathrm{d}t$$

where a and b define the object support for a 1-D object that may be extended into the complex plane by analytic continuation to give

$$F(z = x + iy) = \int_a^b f(t)e^{izt}\mathrm{d}t$$

The Paley–Wiener theorem states that $F(z)$ is an entire function of exponential type which, by the Hadamard factorization theorem, means it can be written as an infinite product of the form

$$F(x + iy) = e^{cz} \prod_{j=-\infty}^{\infty} (1 - (z/z_j))$$

where the z_j's are the locations of the jth complex zero of the function $F(z)$. In this way, a band-limited function of a single complex variable is uniquely defined by its complex zeros. The zeros occur at an infinite number of isolated points in the complex plane, and, in general, these points will not be expected to coincide with the real axis.

To extend the zero description to functions of more than one complex variable, the Plancherel–Polya theorem states that, for an object of compact support, the p-dimensional Fourier transform $F(z_1, z_2, ..., z_p)$ is an entire function of exponential type. $F(z)$ can be uniquely represented by a product of the form

$$F(z_1, z_2) \approx e^{(c_1 z_1 + c_2 z_2)} \prod_{j=1}^{N} \left(1 - \frac{z_1}{g(z_2)} \right)$$

known as the Osgood product shown here for two complex variables and where it is possible that $N = 1$.

Thus, in two or more dimensions, $F(z)$ is represented by the product of either a finite or infinite set of factors. However, *if F is* irreducible, then $N = 1$, and arg(F) is uniquely defined by the magnitude of F. This does not necessarily make that unique phase easy to determine!

A logical consequence of Fourier-based models for fields and waves is the inherent analyticity of the functions being considered. We will address in the next section the practical consequences of having measured noisy discrete data samples, but for now let us focus on these analytic properties. It is well known that the Fourier transform $F(x)$ of a function $f(t)$ will be entire (i.e., analytic throughout the complex $z = x + iy$ complex plane) if $f(t)$ has compact support. $F(x)$ is also known as a band-limited function. In the time domain this simply means that the function is turned on and off, and if t is a spatial variable, then edges or boundaries of some object or aperture will serve the same purpose. Indeed, the presence of just one "edge" or, in the time domain, a causality condition requiring $f(t) = 0$, for $t < 0$ ensures that $F(z)$ is regular in the upper half of the complex z plane. Known as Titchmarsh's theorem, it was pointed out in the 1920s that a logical equivalence to a causality condition is that Cauchy's integral formula can be written for a contour, C, comprising the real x axis and an infinite semicircle in the upper half plane the contribution from which can be neglected. This results in the real and imaginary parts of $F(x)$ being related by Hilbert transformation.

$$\text{Im}[f(x)] = \frac{1}{\pi} \int_{-\infty}^{\infty} \frac{\text{Re}[f(x')]}{x' - x} dx'$$

$$\text{Re}[f(x)] = \frac{1}{\pi} P \int_{-\infty}^{\infty} \frac{\text{Im}[f(x')]}{x' - x} dx' \tag{B.1}$$

These integral transforms allow the real part of F to be computed from the imaginary part and vice versa, and these transforms can be generalized to more than one x dimension. The Kramers–Kronig relations in the temporal frequency domain are logically tied to the causal nature of a medium's polarizability. Interestingly, if one could ensure that $\log F(x) = \log |F(x)| + i\varphi(x)$ were also regular in the upper half of the complex plane, the phase problem would be solved since we could write:

$$\phi(x) = \text{Im}[\log F(x)] = \frac{1}{\pi} P \int\limits_{-\infty}^{\infty} \frac{\text{Re}[F(x')]}{x' - x} dx' \qquad (B.2)$$

Even more interesting is that in 1-D problems, most functions $F(z)$ we encounter have isolated points in the complex plane at which $|F| = 0$. Indeed, much work has been done to study how functions $f(t)$ influence the distribution of these zeros, of which there is an infinite number for a band-limited function. Zeros in the upper half plane lead to singularities inside the contour C, and residues need to be found to correct the phase calculated using Equation B.2. If $F(z)$ is a band-limited function, it can be represented either by its complex amplitude values measured at the Shannon or Nyquist sampling rate, or by the coordinate locations at which these zeros occur.

From $|F|$ measurements, one has only the knowledge of the zero locations to within a sign ambiguity for their imaginary coordinates, y_j. With N complex zeros one can create $\sim 2^{N-1}$ different functions $F(x)$ since F and its complex conjugate have complex zeros that are located symmetrically about the real x-axis. Hence, there are 2^{N-1} functions $f(t)$ that are consistent with the measured $|F|$.

However, if one knew *a priori* that $F(z)$ has a zero free half plane, then Equation B.2 will produce the correct phase. If one applies Equation B.2 regardless, that is, without the prior knowledge of a zero free half plane, then due to the ambiguity in the sign of $|y_j|$, the solution for $f(t)$ that one obtains using the phase calculated from Equation B.2 is the so-called minimum phase function. In 1974, this fact suggested that one simply needs to preprocess a wave prior to detection by adding a known reference wave that satisfied Rouche's theorem and hence ensure that this new superposition of fields represented a minimum phase function. Rouche's theorem states that if function g has N zeros in some region of the complex plane, and function h has $M < N$ zeros in the same region, then if around that boundary, $|h| > |g|$ then $g + h$ will have M zeros in the region. For $h = $ constant, then $M = 0$ providing a mechanism to enforce a zero free region. There is a clear connection here with making a hologram, as mentioned above.

Many iterative phase retrieval methods have emerged over the years and many comparisons between them made. Until a few years ago, the consensus was that several of these methods could be made to succeed but that there remained a degree of uncertainly or an occasional lack of confidence is the results obtained as mentioned above.

The Gerchberg–Saxton algorithm was one of the first such iterative algorithms and it can be described as follows. Given an object function $f(t)$ and its Fourier transform, $F(x)$, the objective of the algorithm is to iteratively recover complex f and F by imposing consistency with measured $|f|$ and $|F|$. The iterations would begin by initially choosing a random phase function followed by reintroduction of the magnitudes after each forward or inverse Fourier transformation, while retaining the current estimate for the phase. The iterations would cease after some noise-determined number of steps or when the changes at each step were less than some acceptable threshold.

The original Gerchberg–Saxton algorithm (Fiddy and Shahid, 2013) was modified by Fienup to use the Fourier modulus and object support and non-negativity constraints on $f(t)$; it became known as "error-reduction" iteration. The nonconvexity of the Fourier magnitude constraint leads to stagnation of this algorithm due to local minima or fixed points that satisfy one constraint but not the other. Stagnation could manifest itself by the presence of the twin images shown earlier, superimposed stripes on the image due to phase wraps near real zeros of F or an inappropriate choice of $p(t)$. These stagnation mechanisms can occur in all phase retrieval methods and various generalizations of these iterative schemes were proposed to avoid it. The most successful modifications have been algorithms where the constraints in the object domain are not rigidly imposed, but are relaxed. For example, taking linear combinations of estimates that are generated can help to avoid stagnation at local minima in the cost function. Fienup developed a very robust approach and developed a family of "input–output" procedures motivated by ideas from control theory. The only limitation of the hybrid input–output map is that the support constraint cannot always be applied, as in crystallography, for example.

REFERENCES

Carter, W. H. 1983. *Inverse Optics: Proc.*, Devaney, A. J. (ed.), SPIE 413, pp. 65–73.
Fiddy, M. A. and Shahid, U. 2013. Legacies of the Gerchberg–Saxton algorithm. *Ultramicroscopy*, *134*, 48–54.

Appendix C: Prior Discrete Fourier Transform

BACKGROUND AND DEFINITION

Since our goal is to recover images of scattering objects with a size similar to the wavelength of the illuminating wave, an image estimate based on an inverse discrete Fourier transformation of the data points in k-space is not sufficient to perform the task. Therefore, we have proposed basing the inverse scattering step on a spectral estimation technique in order to obtain a sufficient resolution and distinguish different types of objects. The method we have been applying is called the prior discrete Fourier transform (PDFT) algorithm. This algorithm had been in use for some time to postprocess back-propagated fields. It is the result of rather recent efforts to adapt the PDFT algorithm as a general substitute for any linear backpropagation procedure.

The PDFT algorithm was developed to estimate signals from sparse discrete measurements, simultaneously incorporating prior knowledge about the object or the measurement system. The PDFT can be applied to a large variety of problems. Typically, measurements provide partial knowledge about the far field of the scattering amplitude. However, in addition, information is available about the typical lateral extension of the object or the area from which the scattered field originates.

For example, in Figure C.1, a low pass filtered image (center) can be enhanced by using the larger outer box in the middle image as the function $p(r)$, to give the image on the right. We assume that the data we get from the scattering experiments is the knowledge of the Fourier space $F(k)$ of some target $f(r)$ at N arbitrarily sampled spatial frequencies k_n:

$$F(k_n) = \int_{-\infty}^{\infty} f(r)\exp(-ik_n r)\mathrm{d}^L r$$

with L referring to the dimension of the problem. We further assume that the target is of compact support and that the object has a physical size, which is described by a real-valued positive function $p(r)$. This prior can be used to incorporate information about the shape as well. In its simplest form $p(r)$ can be a top hat that circumscribes the object entirely. Then, an estimate of the object can be obtained from an inverse DFT

$$\hat{f}_{\mathrm{PDFT}}(r) = p(r)\sum_{n=1}^{N} \hat{F}_n \exp(ik_n r)$$

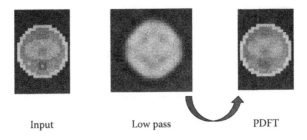

Input   Low pass   PDFT

Figure C.1   Example of low pass filtered images versus a PDFT image.

where $\hat{F}_n$ is determined by the reconstruction algorithm. For instance, to apply an inverse DFT it is assumed that $\hat{F}_n = F_n$. The PDFT is an attempt to find the estimate that is in accordance with the prior information and has a minimum deviation from the real object. Formally, this can be expressed as

$$\chi = \int \frac{1}{p(r)} |f(\boldsymbol{r}) - \hat{f}(\boldsymbol{r})|^2 \, \mathrm{d}^L r \rightarrow \text{minimum}$$

Substituting for $\hat{f}(\boldsymbol{r})$ the $\hat{F}_n$ is the free parameter, and we find the minimum as a solution of a system of linear equations

$$F(\boldsymbol{k}_n) = \sum_{m=1}^{N} \hat{F}_n P(\boldsymbol{k}_n - \boldsymbol{k}_m)$$

with $P(\boldsymbol{k})$ being the Fourier transformation of the prior. The parameter $\hat{F}_n$, which allows calculating the image estimate from an inverse DFT, is obtained as the solution of this system of equations. To solve the system of equations the P-matrix, $P_{n,m} = P(\boldsymbol{k}_n - \boldsymbol{k}_m)$, has to be inverted, which is typically ill-conditioned. To overcome this problem a Tikhonov–Miller regularization is introduced by multiplying the diagonal of the P-matrix with a value $(1 + \varepsilon)$, ε being much smaller than 1. This, in effect, accounts for a homogeneous background noise in the image estimate, and the size of ε necessary to obtain a decent image reflects the signal-to-noise ratio.

We have used the PDFT algorithm primarily to improve the Ewald sphere data coverage as compared with a discrete inverse Fourier transform. We would like to emphasize that the PDFT algorithm can be used to evaluate the quality of data based on knowledge of the target geometry. Then the regularization parameter in combination with the energy of the PDFT reconstruction can be used to quantify the signal-to-noise ratio and detect the presence of multiple reflection artifacts contributing to the data.

Figure C.2 illustrates the performance of the PDFT algorithm. It is evident that the PDFT provides considerable benefit for nonuniformly oversampled data. This is precisely the situation for monostatic backscatter data where only a small number of views is obtained, but with a large number of time frequencies. In this case, the incorporation of prior knowledge frees up the degrees of freedom to improve the image quality and the resolution. It should be noted that the quality of the PDFT reconstruction in Figure C.2 strongly depends

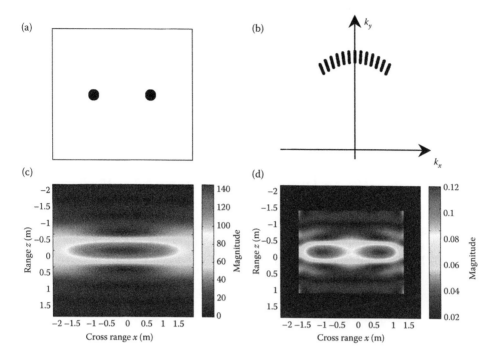

Figure C.2 Demonstration of the PDFT algorithm from synthetic Fourier transform data. (a) Target (two cylinders in free space), (b) map of data in k-space (wide angle ISAR configuration), (c) DFT estimate of target, and (d) PDFT estimate.

on the signal-to-noise ratio. In other words, the PDFT is capable of providing the optimum linear image estimate based on a given set of data points, prior information about the target, and the quality of the data.

The Power of the PDFT

From a mathematical perspective, the PDFT is thus a method for estimating an unknown vector or a function (an object such as a discrete image, for example) from linear functional information about that vector that by itself is insufficient to specify uniquely the vector in question. In such cases, it is crucial to use prior information to single out one specific estimate from the many possibilities. The approach we use with the PDFT is to select *a priori* information about the target and incorporate it into the design the PDFT estimate. It is a method for reconstructing (estimating) an object from many finite linear functional values (linear "projections," if you will). The PDFT is designed to be applied to the underdetermined problem, to combat the degrading effects of limited data and nonuniqueness of solutions through the inclusion of prior information about the object to be reconstructed. In simplest terms, the data is insufficient to determine a unique solution. The significance of this approach is that it permits us to incorporate prior information about the object to be recovered. We illustrate that point now using the case of reconstruction for Fourier transform data.

Suppose now that our object $f = f(x)$ is a function of a continuous real variable x in R. We use f to denote either V_u, V, or a target region within V, depending on the problem we are facing. The Fourier transform of f is $F(\omega)$, the function of the continuous real variable ω in R given by

$$F(\omega) = \int f(x)\exp(-ix\omega)dx$$

where the integral is over the real space R. F here denotes the far field arising from the domain D enclosing V. Let our data be many finite values of the Fourier transform of f; that is, let the data be

$$d_n = F(\omega_n) = \int f(x)\overline{\exp(i\omega_n x)}dx, \quad \text{for } n = 1,2,...,N$$

where ω_n's ($n = 1,2,\ ...,N$) are N arbitrary points in R. Let $p(x) \geq 0$ be a prior estimate of the profile of the function $f(x)$ to be reconstructed. Let H consist of all linear combinations of functions of the form $h_m(x) = p(x)\exp(i\omega_m x)$, $m = 1,2,...,\ N$. We then have

$$d_n = \int h_m(x)\exp(-i\omega x)dx = \int h_m(x)\exp(-i(\omega_m - \omega_n)x)dx$$

Therefore, $d_{mn} = P(\omega_n - \omega_m)$ for each m and n, where P is the Fourier transform of p. The vector $b(x)$ has entries $p(x)\exp(i\omega_m x)$. With $e(x) = (\exp(i\omega_1 x)\cdots\exp(i\omega_N x))^T$ we have $b(x) = p(x)e(x)$. Then the PDFT estimate of $f(x)$ is

$$\hat{f}(x) = a(x)^T d = b(x)^T D^{-T} d = p(x)e(x)^T c$$

where $c = D^{-T}d$.

Two important observations can now be made. The PDFT estimator is easily regularized in the Tikhonov–Miller sense when data are noisy. Secondly, the prior constraints employed can be very general. For example, one could use the radar beam pattern as prior knowledge of where the scattering arises from, if one is processing the radar returns directly. If an image has been formed, for example, as a SAR image, then one can apply a window around the region of interest and improve the resolution within that subdomain. The effectiveness of this depends on the proportion of energy in the cluttered background to that in the vicinity of the target. A smaller window resolves this problem and actually assists with the resolution enhancement step, as is clear from the theoretical section above. An example is shown here of real data inversion imaging a calibration sphere and a smaller plastic landmine (Figure C.3).

We note that a tomographic reconstruction can alleviate problems of the strong return, which comes from the surface of the ground, a particular problem for mine and bunker detection. The ease with which this can be done depends on the point spread function or side lobes arising from the limited data. The PDFT can improve the point spread function, but it can also assist with this problem before the image formation step. Through appropriate k-space gating one can remove the data most strongly associated with a surface reflection prior to using the PDFT, as shown in Figure C.4.

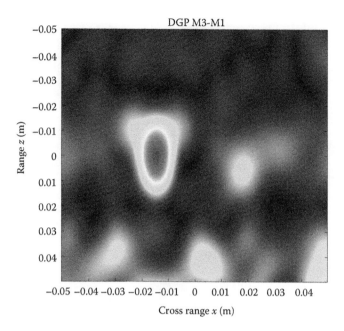

Figure C.3 Imaging buried objects from scaled ISAR experiment (calibration metal sphere and VS-50 plastic landmine on right). The magnitude is reconstructed from the difference of the ground plate with and without objects present.

ALGORITHMS FOR TARGET IDENTIFICATION

Based on the processing steps described above, an estimate of the complex permittivity distribution in the 3-D domain of interest (either V or a sub-domain within V) will be available. Further processing extracts either an image or signature to identify the target. This step could come before or after a need to track a suspected target. It is also at this point that multiple images, taken with two or more radar frequency bands or subapertures, would be used in conjunction with each other. A goal of this stage is to unambiguously discriminate between decoys, camouflaged structures and nontargets in order to achieve high recognition rates and low false alarm rates. So, having improved the resolution using prior knowledge, the resulting image or tomographic reconstruction of the image domain can be further processed to extract specific target information. In the case of SAR imagery, the initial image resolution is a function of the size of the radiating antenna. In strip scan SAR as opposed to spotlight mode, the volume V being irradiated is constantly changing as it moves through the radar footprint. This requires a modified model for the PDFT to be considered in which a 4-D k-space must be considered, the fourth dimension representing a weighting applied to the 3-D k-space associated with the sequence of domains V from which scattered field data are collected.

A Statistical Approach

The PDFT estimates a target such as a discrete image using prior information to single out one specific estimate from the many possibilities. The approach we use here with the PDFT is to select a parametrized family of objects

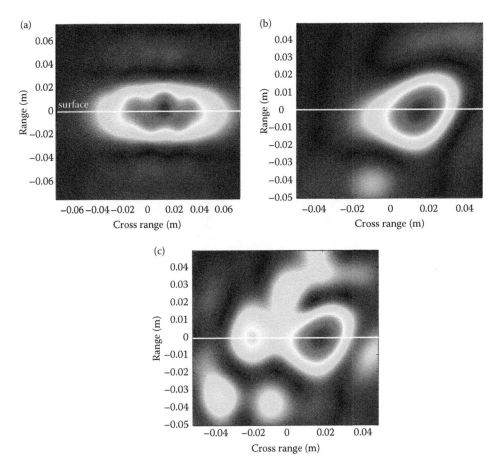

Figure C.4   Imaging of buried objects (magnitude). (a) DFT estimate of two metal spheres (angle range 0–60°), (b) DFT estimate for angle range 14–50°, and (c) PDFT estimate for angle range 14–50°. Limiting the angular range of the data (from (a) to (b)) suppresses the surface return and then the PDFT improves resolution (in (c)).

a priori and design the PDFT estimate to perform well on the members of this family. If the family incorporates our prior information about the object to be recovered and this information is reasonably sound, then the PDFT estimate based on the actual data should be good. If there are no more parameters than there are data points, then the PDFT is designed to estimate the members of the family exactly; if not, then the PDFT performs well, on average, across the members of the family. Random noise can be included through the use of an infinitely parameterized family. As the data is processed and new information is obtained about the object of interest, the family can be modified to incorporate this information and the algorithm restarted, providing a data adaptive extension of the PDFT. The idea of designing an estimation procedure by requiring it to perform well on a family of possibilities is used in numerical quadrature and lies at the heart of statistical estimation, where the family is usually called an *ensemble*.

The PDFT philosophy is to use our prior knowledge to design a parametrized family of possible solutions and then to design an estimation procedure that is exactly correct (if possible) for the members of the family. If the object to be recovered is reasonably well approximated by members of this family, the estimation procedure so constructed will work well on the actual data we have measured. The PDFT can therefore be viewed as a (somewhat generalized) statistical procedure, in the sense that the method is optimized for a family (an *ensemble*) of possibilities. Similar methods are used in numerical quadrature, in which one attempts to estimate the integral of a function from sampled data; estimates are calculated to be exact on a family, say, of polynomials, and then applied to the measured data.

In the discussion above describing the PDFT, we assumed that the family H is an N-dimensional linear subspace; assuming that the matrix D is invertible, we solve the matrix equation $b(x) = Da(x)$ for $a(x)$. When H is a more complicated family, perhaps having no finite parameterization, we seek a least squares solution of $b(x) = Da(x)$. It can happen that the vector $b(x)$ has one entry for each member of the family H making it an infinite dimensional vector. We then consider $D^T b(x) = D^T Da(x)$; the vector $D^T b(x)$ involves a sum over all the members of H, but the result $D^T b(x)$ is an N-dimensional vector. In this case, we would model $D^T b(x)$ and $D^T D$ directly, rather than performing the infinite summation.

A standard problem in numerical analysis is the estimation of a definite integral, say $I = \int_a^b g(t)dt$, from finitely many values of the function g, say $g(t_1),\dots,g(t_N)$. The usual approach is to take as our estimate of I a linear combination of the data values; that is, let our estimate be $\hat{I}$ given by

$$\hat{I} = \sum_{n=1}^{N} a_n g(t_n)$$

The next step is to determine what the values of the coefficients $a_1,\dots,a_N$ should be. One way to determine these coefficients is to select them so that the estimate $\hat{I}$ is exactly equal to I for some set of special functions, g, such as polynomials of degree $N-1$ or less. Now let us apply this philosophy to the Fourier transform estimation problem.

Our problem now is to estimate

$$f(x) = \frac{1}{2\pi} \int_{-\infty}^{\infty} F(\omega) \exp(i\omega x)dx$$

from the data $F(\omega_1),\dots,F(\omega_N)$. For each fixed value of x this is an integral estimation problem. So for each fixed value of x, we determine a set of coefficients that depend on x, say $a_1(x),\dots,a_N(x)$, and take as our estimate of $f(x)$ the quantity

$$f(x) = \sum_{n=1}^{N} a_n(x) F(\omega_n)$$

The next step is to select a special class of functions for which this estimation procedure must work perfectly and determine the coefficients needed to have that.

We suppose that we have $p(x) \geq 0$ as our prior estimate of the overall shape of the function $f(x)$; let $P(\omega)$ be the Fourier transform of $p(x)$. Let us take as the special functions the set $H_m(\omega)$, with $m = 1, \ldots, N$, where $H_m(\omega)$ is the function whose inverse Fourier transform is $h_m(x) = p(x)\exp(ix\omega_m)$; that is, $H_m(\omega) = P(\omega - \omega_m)$. Now we apply our estimation procedure to each of these $H_m(\omega)$.

We fix a value of m and apply the estimation procedure in the previous equation to estimate $h_m(x)$. We then have

$$h_m(x) = \sum_{n=1}^{N} a_n H_m(\omega_n)$$

for each $m = 1, \ldots, N$. Putting in what H_m and h_m are, we obtain the system of equations

$$p(x)\exp(ix\omega_m) = \sum_{n=1}^{N} a_n(x) P(\omega_n - \omega_m)$$

for $m = 1, \ldots, N$.

Let D be the square matrix with entries $P(\omega_n - \omega_m)$, let $a(x)$ be the column vector with entries $a_n(x)$, and let $e(x)$ be the column vector with entries $\exp(ix\omega_m)$. We can then write the previous equation as $p(x)e(x) = Da(x)$. Solving for $a(x)$, we get

$$a(x) = p(x)D^{-1}e(x)$$

We now apply these coefficients to the original Fourier transform estimation problem. Let d be the column vector with entries $F(\omega_n)$. Our estimate of $f(x)$ is then

$$\hat{f}(x) = a(x)^{\mathrm{T}} d = p(x)e(x)^T D^{-T} d$$

which is the PDFT.

Morphing the Prior

The resolution of the PDFT estimate systematically improves as the prior $p(r)$ improves. When no assumptions are warranted regarding the prior function, one can then choose a generic prior and then reduce its size and shape and monitor the energy of the associated PDFT estimate. It can be shown that as the prior shrinks to a size smaller than the target shape, the energy of the PDFT estimate dramatically increases. A systematic approach to this "morphing" of $p(r)$ allows target information to be deduced, either an improved image of the target or a defining signature shape (Figure C.5). This is illustrated below.

By identifying a region in an image that looks like a likely target, one can extract this array of pixels (a large enough area around the target which includes the significant portion of the SAR point spread function), and then

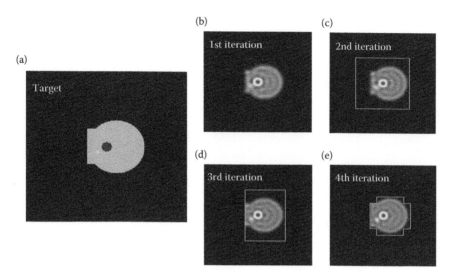

Figure C.5 The original target in (a) is reconstructed from limited data using no prior information in (b) followed by priors that encroach on, that is, morph to the exterior shape of the target.

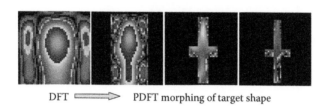

DFT ⟹ PDFT morphing of target shape

Figure C.6 Image of a missile target using DFT and then having the PDFT morph technique applied.

apply the PDFT to this subimage. This is illustrated with the simulated target below.

This is also illustrated in the example below using real data taken from a model missile provided by US Air Force Research Laboratory (AFRL).

In this example, data gathered from a model missile were used to form an image as shown on the left (Figure C.6). Only 0.5% of the data or 16×16 k-space data were used to calculate this and no details of the missile can be seen. Employing the PDFT and "morphing" the prior information representing a dynamic perimeter shape, one can shrink this prior and monitor the energy associated with the corresponding estimate. This approach is very sensitive to prior shapes that impinge on actual missile pixels, allowing us to define an outline as shown on the right. When too little data are gathered to form a reasonable image, this approach allows a signature outline to be determined which is characteristic of the target. We believe that this provides a unique classifier for different targets.

Appendix D: The Poynting Vector

The Poynting vector relates the movement of electromagnetic power to the temporal variation of stored energy, that is, the magnitude and direction of power flow. It is defined to be $\bar{S} = \bar{E} \times \bar{H}$ and at high frequencies, the time averaged Poynting vector can be shown to be $\bar{S} = (1/2)\text{Re}[\bar{E} \times \bar{H}^*]$ where Re denotes the real part. This simple form assumes that $E \propto H$, that is, that the fields are simply proportional to each other and hence the medium is isotropic. If the medium is such that its refractive index $n < 0$, then $\bar{S}$ is necessarily in the direction opposite to $\bar{k}$. We normally assume a plane wave solution to the wave equation and assume that $\bar{H}$ is perpendicular to $\bar{E}$ and $\bar{k}$. In free space, $H = (E/\eta_0)$, and $\eta_0 = (\mu_0/\varepsilon_0) = 120\Omega$. By convention, $\bar{E}$, $\bar{H}$, and $\bar{k}$ form a right-handed set (Figure D.1). In some medium, the phase velocity of the wave is given by $(c/\sqrt{\mu_r\varepsilon_r}) = (\omega/k)$ and $H = (E/\eta)$, where $\eta = \eta_0\sqrt{\mu_r/\varepsilon_r}$.

Let us consider the boundary conditions (Figure D.2). The quantities shown in this figure are continuous; here t denotes the tangential component and n the normal component. If ε_{eff} is for a metal, then $E_{2t} = 0$, but we may have surface waves and surface currents, and surface plasmons to contend with. These boundary conditions are

$$E_{1t} - E_{2t} = 0$$

$$B_{1n} - B_{2n} = 0$$

$$D_{1n} - D_{2n} = \rho_s \ \text{(surface charge)}$$

$$H_{1t} - H_{2t} = \bar{J}_s \times \hat{n} \ \text{(surface current)}$$

Another important point concerns the energy density in a medium which is given by

$$W = \frac{1}{2}\varepsilon(\omega_0) |E|^2 + \frac{1}{2}\mu(\omega_0) |H|^2$$

Otherwise, assuming little dispersion near ω_0 we get

$$W = \frac{1}{2}\frac{d(\omega\varepsilon(\omega))}{d\omega}\bigg|_{\omega=\omega_0} |E|^2 + \frac{1}{2}\frac{d(\omega\mu(\omega))}{d\omega}\bigg|_{\omega=\omega_0} |H|^2$$

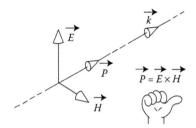

Figure D.1   Illustration of the right-hand rule.

Figure D.2   Illustration of boundary conditions.

Appendix E: Resolution and Degrees of Freedom

We gain some additional understanding of the inverse scattering problem and its challenges by considering the scattered field measurement process, regarded as Fourier data, as a truncated sampling process. For a finite object of width D the Whittaker–Shannon sampling theorem demands a sampling rate of at least $B_{min} = 1/D$ in the frequency spectrum. The representation of the k-space spectrum is only complete, however, if an infinite number of samples are available covering the entire k-space. Since our k-space volume is physically limited to include only propagating waves, it is necessarily a truncated set of samples.

The image estimate based on a truncated set of k-space samples provides only a low spatial frequency estimate of the signal. For a spectrum sampled at the Nyquist rate determined by the object support, we can easily verify that the P-matrix of the PDFTs is diagonal. This means that at this sampling rate the PDFT will not improve the image estimate beyond the classical limit, and the model of the signal is already represented optimally in a least square sense by the available truncated sampling expansion, that is, by a discrete Fourier series.

It is worth contemplating why Fourier data sampled at the Nyquist rate precludes any hope of bandwidth extrapolation. The sampling expansion is constructed to obtain orthogonal interpolation functions. In other words, the zeros of the interpolating *sinc* function perfectly coincide with the location of sampling points. Thus, the data are assumed to be independent and do not contain any information related to other sampling points. Points on the sampling grid of the spectrum outside the window of measured data do not contribute to the points inside this window. In turn, the latter cannot be used to estimate points on the sampling grid outside the data window. It is well known that spectral data sampled at the Nyquist rate do not contain sufficient information for bandwidth extrapolation but that a higher sampling rate is required. This is precisely the context to which the PDFT algorithm is applicable. For oversampled data, the samples are no longer independent, and the coefficients of the PDFT reconstruction must be selected to ensure data consistency as a result of convolution with the interpolating function. This interdependency has two consequences. First, it provides the freedom to balance the PDFT coefficients to obtain improved signal resolution (and bandwidth extrapolation). Second, the image reconstruction from interdependent samples results in high susceptibility to noise in the measured data. In particular, the improved image resolution is the result of a delicate interference between different interpolating functions, all of which carry the main portion of their energy outside the data window. Thus, even small errors inside the window are amplified in the extrapolated region. It has been shown that this confines the accurately extrapolated frequency band to a rather small domain outside the known data window. While we may argue that further increasing the rate of sampling might balance the lack of accuracy, this is not the case; the only way to extend the extrapolated region is to improve the data quality.

It was suggested and later verified that this instability can be interpreted as being linked to the formation of super-oscillations. In others words, numerical super-resolution algorithms construct super-oscillating signals, which optimally resemble the object function. We note that this relationship between super-resolution algorithms and optical super-resolution phenomena was exploited from a practical point of view by constructing super-resolving filters with the PDFT algorithm.

The strong dependence of the achievable bandwidth extrapolation on the SNR of the data also points to a trade-off in the Lukosz sense. The space-bandwidth product of the measured signal can be calculated straightforwardly as $S = B_{max}D$ and we attempt to extract information about more than S independent signal features by trading for SNR. This is the same relationship we found for super-resolving filters and super-oscillations. The exponential growth of the required SNR as a function of bandwidth can be identified as the main constraint for super-resolution imaging.

The interpretation of the PDFT algorithm as a "Lukosz trade-off" leaves a pessimistic outlook on the prospect of calculating super-resolved images. However, it has been observed that many numerical super-resolution algorithms report significant improvements over the classical diffraction limit. They account for this by distinguishing between what they call primary and secondary super-resolution. We interpret primary super-resolution as the gain in resolution due to exploiting the Lukosz trade-off between signal-to-noise ratio (SNR) and image resolution. This gain in resolution is essentially independent of the object signal and the number of samples and adheres to the properties of the PDFT algorithm discussed so far. Secondary super-resolution is defined as any additional gain in image resolution not related to primary super-resolution. This gain may be related to other Lukosz trade-offs, particularly in the context of image reconstruction from multiple encoded image frames. For instance, imaging through atmospheric turbulence and image reconstruction from multiple aliased images may be interpreted as extracting the desired information simply from multiple separate measurements. However, in each frame the different frequency bands overlap and the information gain is transmitted similar to the degrees of freedom accessible through primary super-resolution. It is not surprising therefore that similar algorithms are used to recover the image and that in these cases the algorithm displays superior image resolution simply based on the better SNR of the high-frequency components.

However, a second source of secondary super-resolution is due to the use of prior information, and the PDFT algorithm can be used to understand this type of secondary super-resolution more intuitively as an interrelationship between the estimator and the object function. In particular, we can characterize the estimator in terms of the prior knowledge we inject into the reconstruction process. Since we strive to estimate a continuous signal from a finite set of data, we try to solve an ill-posed problem and always must make assumptions to select the true object function from the infinite set of possible signal reconstructions. We interpret secondary super-resolution as the result of incorporating specific object characteristics into the choice of prior information. It is immediately obvious that the PDFT algorithm introduces additional information through the choice of prior and the *P*-matrix. These are composed of the weighting function $p(x)$ and the regularization parameter, ε, the latter in effect improving the estimation process by accounting for signal noise. The trade-off between the amount of data and the use of prior information can be further emphasized if we consider the trivial case $|p(x)|^2 = V_{obj}(x)$ (i.e., the prior is chosen to be the true object function). The reconstruction is always perfectly accomplished from a single sample in the Fourier plane. It is clear that the choice of prior may incorporate object properties to any degree, and a weighting function $p(x)$ already consisting of small features to some degree facilitates the extrapolation of the spectrum in the Fourier domain. Conversely, we may accomplish an improved regularization by using a smoothly modulated weighting function.

Such estimators have indeed proved superior for resolving localized object structures and may be interpreted in terms of having selected a more appropriate basis set for the given object being imaged. This is why some *a priori* knowledge of a support constraint for a compact scatterer is so powerful. Effectively, the system degrees of freedom are put to much better use!

Appendix F*: MATLAB® Exercises with COMSOL® Data

As a supplement to the materials in this book, the MATLAB® and COMSOL® data files used to create most of the images presented are provided to the reader for demonstration, experimentation, and further development. This COMSOL data was generated using the method described in Chapter 5. The MATLAB code incorporates many of the theory and techniques discussed throughout this book. The main program has been broken down into smaller programs to help focus on different aspects of the code. The first program is called **Ewald** and is used to create 2-D Ewald diagrams from the COMSOL data files. The second is the program called **Born** which is used to generate first Born approximation reconstructions using the provided COMSOL data files. The third is the program called **Cepstrum**, which is used as the next step in the imaging process of applying a cepstrum filter to the reconstructed image files. An additional version of this program is **Cepstrum2**, which is the same as **Cepstrum** but adds an additional reconstruction where the incident wave is subtracted in cepstrum space for improved performance. The final program is called PDFT, which is used to demonstrate the application of the Prior Discrete Fourier Transform (PDFT) as described in Chapter 7 and Appendix C.

The following suggested exercises are included to help the reader use and understand these codes and to see how they relate to the information presented in the text. In addition to the suggested scenarios that could be tried, some direction is given for the assessment and interpretation of the results. These exercises were developed to be used as exercises in a college course, but are also helpful to the general reader in understanding the concepts discussed in the book, and to understand the MATLAB code and how it works.

Ewald Exercises

1. Use **Ewald** in MATLAB to generate Ewald circles for various targets. Do this several times varying the target, the permittivity, the number of sources, and the number of receivers. In all cases, where does the majority of useful data seem to reside? What does this suggest? If the opposite were true, what might this suggest? What might do a better job of filling the space, increasing the number of sources or the number of receivers? How would the images change if the incident frequency were increased or decreased? Why?

---

* Supplementary materials including MATLAB code for exercises are available on the book's page at www.crcpress.com. Please visit the site, look up the book and click to the Downloads and Updates tab.

2. Use **Ewald** in MATLAB to generate Ewald circles for a target consisting of one circle (Choice 0) and two circles (Choice 1). Use any valid permittivity and a minimum of 12 sources and 120 receivers. For each target save a 2-D output image of the combined Ewald circles. How are these images similar? How are these images different? Please discuss your observations and speculate on reasons for both the similarities and differences.

3. Repeat the procedure in Exercise 1 for one square (Choice 2) and two squares (Choice 3). Compare the resulting images obtained here with the images obtained in Exercise 1. Discuss similarities and differences and reasons for them.

4. Use **Ewald** in MATLAB to generate Ewald circles for two triangles (Choice 5) using 4 sources and 360 receivers. Do this twice, once for a target permittivity of 1.1 and again for a target permittivity of 1.9. Be sure and show the individual Ewald circles for each source. For each scenario save the individual Ewald circle for Source #2. Compare the images for each Ewald circle. Describe the differences and discuss why you think they are so different.

Born Exercises

1. Use **Born** in MATLAB to generate Born approximation reconstructions for various targets. Do this several times varying the target, the permittivity, the number of sources, and the number of receivers. Discuss your observations for the various scenarios and give some speculation for reasons for the variations observed.

2. Use **Born** in MATLAB to generate Born approximation reconstructions for a circle (Choice 0) using 36 sources and 360 receivers. Also, set the image maximum dimension to 0.25 for each case. Do this three times for a target permittivity of 1.1, 1.5, and 1.9. For each scenario save the Born images generated. Compare the images for each Born reconstruction and discuss the differences. Please speculate as to the reason for these differences.

3. Use **Born** in MATLAB to generate Born approximation reconstructions for any target consisting of two objects (Choices 1, 3, or 5) using any valid permittivity and a minimum of 6 sources and 120 receivers. Be sure and show the individual Born reconstructions for each source. Please discuss your observations for the individual Born reconstructions for each source and speculate on their relationship to the Born of the combined sources.

4. Use **Born** in MATLAB to generate Born approximation reconstructions for two circles (Choice 1) using a minimum of 12 sources and 120 receivers. Do this twice, once for a target permittivity of 1.1 and again for 1.9. For each scenario save the individual Born reconstruction for Source #12. Compare the images for each reconstruction. Describe the differences and discuss why you think they are so different.

Cepstrum Exercises

1. Use **Cepstrum** in MATLAB to generate Born and Cepstrum reconstructions for various targets. Do this several times varying the target, the permittivity, the number of sources, and the number of

receivers. Discuss your observations for the various scenarios and give some speculation for reasons for the variations observed. Which method performs best for image shape? Which method performs best for proper scale? Please speculate as to why you think this is the case.

2. Use **Cepstrum** in MATLAB to generate Born and Cepstrum reconstructions for a two squares (Choice 3) using 12 sources and 120 receivers. Use any valid permittivity. Inspect the images for the presence of aliasing. Do this as many times as necessary, decreasing the number of receivers each time, until aliasing is present in the images. What is the number of receivers when aliasing is first observed? Use the method (and receiver distance) discussed in the book to calculate the maximum spacing that should be to avoid aliasing. How does this compare with what you observed experimentally? Now re-run the same scenario where the aliasing was first observed, except this time set the jitter bandwidth to 2 and inspect new image for aliasing. Do this one more time except set the jitter bandwidth to 4 and again inspect the resulting image. Does this eliminate the aliasing from the reconstructed image? What is the impact on the Ewald circles?

3. Use **Cepstrum** in MATLAB to generate Born and Cepstrum reconstructions for any target consisting of two objects (Choice 1, 3, or 5) using any valid permittivity and a minimum of 6 sources and 120 receivers. Be sure and show the individual Born and Cepstrum reconstructions for each source. Please discuss your observations for the individual Born and Cepstrum reconstructions for each source and speculate on their relationship to the Born and Cepstrum of the combined sources. Which method performs the best? Why do you think this is?

4. Use **Cepstrum** in MATLAB to generate Born and Cepstrum reconstructions for a combination of one circle, one square, and one triangle (Choice 6) using a minimum of 12 sources and 120 receivers. Set the permittivity equal to 1.2. Run this scenario three separate times with everything identical with the exception of varying the filter sigma value for sigma equal to 1, 5 (default), and 50. Observe and compare the effects of varying sigma on each reconstruction method. Does it affect each method the same? Why or why not? What seems to be the optimal value for sigma for each case?

5. Use **Cepstrum** in MATLAB to generate Born and Cepstrum reconstructions for a combination of one circle, one square, and one triangle (Choice 6) using a minimum of 12 sources and 120 receivers. Set the permittivity equal to 1.2. Run this scenario three separate times with everything identical with the exception of varying the filter multiplier value to 0.01, 0.0555556 (default), and 1. Observe and compare the effects of varying the multiplier on each reconstruction method. Does it affect each method the same? Why or why not? What seems to be the optimal value for this multiplier for each case?

6. Use **Cepstrum** in MATLAB to generate Born and Cepstrum reconstructions for a combination of one circle, one square, and one triangle (Choice 6) using a minimum of 12 sources and 120 receivers. Set the permittivity equal to 1.2. Run this scenario three separate times with everything identical with the exception of varying the Reference

Multiplier value to 0.1, 1 (default), and 10. Observe and compare the effects of varying the multiplier on each reconstruction method. Does it affect each method the same? Why or why not? What seems to be the optimal value for this multiplier for each case?

PDFT Exercises

1. Use **PDFT** in MATLAB to generate PDFT, Born and Cepstrum reconstructions for a square target (Choice 2) using a permittivity of 1.2 and a minimum of 12 sources and 120 receivers. For the prior use a square shape with the default widths. (Set maximum image scale to 0.5 m for better results.) How does the reconstructed image using the PDFT method compare with the Born reconstruction? Which method performs the best? Why do you think this is?

2. Repeat the process described in the previous example applying the PDFT method described, and then once again without using the PDFT method. This time compare the performance of the Cepstrum method outputs for each case. How do these reconstructed images compare? Discuss your observations for each Cepstrum reconstruction paying special attention to the scales. Which method performs the best? Why do you think this is?

3. Repeat the process described in the previous example applying the PDFT method described, with the exception of setting the prior widths equal to 0.06. Please discuss your observations for the PDFT and the Cepstrum reconstructions. How does this affect the results? Why do you think this is?

4. Repeat the process described in the previous example applying the PDFT method described, with the exception of setting the prior offsets equal to 0.1. Discuss your observations for the PDFT and the Cepstrum reconstructions. How does this affect the results? Why do you think this is?

5. Use **PDFT** in MATLAB to generate PDFT, Born, and Cepstrum reconstructions for the square and circle target (Choice 4) using a permittivity of 1.2 and a minimum of 12 sources and 120 receivers. For the prior, use a two shape combination of a circle and a square that are centered on the appropriate corresponding target location. The prior shape sizes should be slightly bigger than the target dimensions. Save images of the reconstructed images and the prior shapes for review. Did the use of the prior improve the reconstructed images?

6. Use **PDFT** in MATLAB to generate PDFT, Born, and Cepstrum reconstructions for any target consisting of two objects (Choice 1, 3, or 5) using any valid permittivity and a minimum of 6 sources and 120 receivers. Run the reconstructions twice, using PDFTs as input to the Cepstrum method and again using only the Born as inputs to the Cepstrum method. Be sure to show the individual Born and Cepstrum reconstructions for each source. Discuss your observations for the individual Born and Cepstrum reconstructions for each source for both methods. What are some of the differences between using the two inputs? Which method performs the best? Why do you think this is?

7. Use **PDFT** in MATLAB to generate PDFT, Born, and Cepstrum reconstructions for the target set consisting of a triangle, a square, and a circle

(Choice 6) using a permittivity of 1.2 and a minimum of 12 sources and 120 receivers. For the prior use 3 separate prior shapes to match the target configuration (some experimentation may be required). How does the reconstructed image using the PDFT method compare with the Born reconstruction? Once the prior was properly configured did it improve the performance of the reconstructed images?

Advanced Exercises

1. Use **Cepstrum2** in MATLAB to generate Born and Cepstrum reconstructions for any target consisting of two objects (Choice 1, 3, or 5) using a permittivity of 1.5 and a minimum of 12 sources and 120 receivers. Please discuss your observations for each Born and Cepstrum reconstruction paying special attention to the scales. Which method performs the best? Why do you think this is?

2. Choose any target and any valid permittivity available in **Cepstrum**. Using your choice, calculate the 2-D value for N, the minimum degrees of freedom number described in Chapter 6. Now using your scenario choice and the associated calculated value of N, generate reconstructions using **Cepstrum** such that the combined choice for number of sources and receivers satisfies the conditions for N, $N^2/2$, N^2, $2N^2$, $4N^2$, $8N^2$, and $16N^2$. Do not forget to add jitter bandwidth as needed to overcome aliasing. For which scenario does the reconstructed image begin to look useful? What happens to the image as you increase the source/receiver numbers above this threshold?

3. Repeat Exercise 2 above except this time use **PDFT** in lieu of **Cepstrum2**. Utilize the PDFT method in your reconstructions and see if this has any effect on the performance of the reconstructions. Does this change the minimum threshold for degrees of freedom?

Index

Printed and bound by CPI Group (UK) Ltd, Croydon, CR0 4YY

01/11/2024

01782601-0001